IMMORTALITY:

THE SCIENCE OF FORBIDDEN FRUIT

by

Michael A. Tewell, MA, JD

Copyright 2019

All rights Reserved

PUBLISHED 2019

IMMORTALITY: THE SCIENCE OF FORBIDDEN FRUIT; copyright 2019 by Michael A. Tewell. All Rights Reserved. No part of this publication may be reproduced, stored, or transmitted in any form by any means, including but not limited to digital, electronic, mechanical, photocopying, recording or otherwise; or conveyed via the Internet or any website or other media without the prior, written permission of the author; except in the case of brief quotations embodied in news articles, reviews, academic research and books (electronic and printed).

Library of Congress Cataloging-in-Publication Data

Tewell, Michael A.

IMMORTALITY: THE SCIENCE OF FORBIDDEN FRUIT / Michael A. Tewell / a work of non-fiction which includes referenced citations and illustrations.

ISBN: 978-0-9600607-1-9 (paperback)

ISBN: 978-0-9600607-2-6 (future hardcover)

ISBN: 978-0-9600607-0-2 (ebook)

1. Christianity. 2. Evolution. 3. Genetics. 4. Immortality. 5. Individuality. 6. Neuroscience. 7. Quantum Mechanics. 8. Religion and Science.

Cover Image: SelfPubBookCover.com / LadyLight

Cover Design: SelfPubBookCover.com / LadyLight

Inquiries should be addressed to the author: Michael A. Tewell;

P. O. Box 1314; Palm Harbor, Fl. 34683; or, at michaelatewell@gmail.com

PUBLISHED IN THE UNITED STATES OF AMERICA

CONTENTS

Contents..Page 3

Acknowledgements...Page 5

Introduction..Page 6

PART ONE: THE THEORY: ... .Page 9

1. The Crossroads… ....Page 10

2. Fake News.....................................Page 20

3. PhiAlpha:The Seedof Life......................Page 34

4. Evolution: The Quiet Creator..................Page 59

5. Individuality: Child of Our Memories..........Page 79

6. PhiAlpha Activation...........................Page 90

7. Memory Organization...........................Page 110

8. Memory Prioritization.........................Page 119

9. Memory Encoding...............................Page 131

10. Memory Emission……………………………….Page 139

11. Memory Reception……………………………..Page 147

12. Validation Testing……………………………...Page 158

13. Conclusions…………………………………….Page 171

PART TWO: PHIALPHA…………...……………....Page 178

14. PhiAlpha…………………………………….....Page 179

15. Dying Day……………………………….……..Page 193

16. Camelback Mountain…………………………..Page 216

17. Walking Two Paths…………………………….Page 253

18. The Dream……..………………………………..Page 268

19. The Quantum Sea………………………………..Page 281

20. The Next Step…………………………………..Page 287

References……………………………………..Page 296

Illustrations……..……………………………….Page 311

The Author……………………………………..Page 314

ACKNOWLEDGEMENTS

No one has sacrificed more as I spent hours researching and writing this book than my wonderful wife, Clementine "CC" Conde, and our own bright and shining star of a teenage son, Alex. Both had their own lives to live, yet they willingly put up with me as I wrestled this dragon of a book to completion. They are the best and I am forever grateful for their patience.

The author extends his appreciation to Creative Common licensees, who have allowed me to share their illustrations in this book. Certain parts of the theoretical model are technically worded, and the illustrations Creative Commons has allowed me to use are a valuable visual aid to help the reader understand the processes involved in PhiAlpha.

Finally, I am greatly indebted to my copyeditor, Sharon Kaplan of editingmatters1@gmail.com, for her extraordinary skill and professionalism. She has been a wizard with how she guided me painlessly through the treacherous landscape of manuscript editing. Without her steady hand, this book does not get done.

INTRODUCTION

Forty years ago, I chose to spend my professional life as a trial lawyer rather than as a Christian priest. The reason for this choice was the result of a conflict between my Christian upbringing and my personal life experiences that began as the result of a near-death experience (NDE) I had as a young child.

My near-death-experience at the age of 6 years old was both a gift and a curse. My NDE gave me a certain knowledge of the force that guides us through our lives. But, this experience also made me question how we humans interpret that force. My NDE made me wonder whether there wasn't an alternative to the Christian belief of a heaven, hell, and an afterlife A tiny crack was created in the brainwashing we all receive as children when taught our parents' religious beliefs.

But I was only a child. I had no scientific training. I was only an elementary school student. I had no idea how to prove or disprove such a concept of an alternative to the Christian afterlife. I was too intimidated by the system of religious brainwashing in our society to even raise the issue with my middle-class Midwestern parents. They wanted me to be a priest. Or, at least, an honest plumber. I kept my questions to myself. I knew I would have to resolve them on my own.

As I grew older, this crack grew into a chasm separating me from the Christian faith I had been raised in. Simply put, what I was being taught on Sunday mornings in church and Bible school did not square with what I experienced in my day-to-day life.

This conflict had a direct impact upon my career choice. I wanted to share my spiritual knowledge with others, but no matter how much I wanted to be a priest to make my parents happy, I could not justify being a spiritual hypocrite. How could I be a priest when my life experience taught me the Christian view of heaven and hell and the afterlife is a lie?

My decision to become a lawyer allowed me to live my life without having to lie about my spiritual beliefs. But I found that, being a lawyer, I would have to postpone my analysis of whether there is an afterlife as Christians claim.

During my 25 five years as an assistant public defender, I had no time to devote to the questions and issues that so impacted my youth.

The law became a priesthood all its own to me. The courtroom became my personal temple, jury trials my signature mode of self-expression. I reveled in the experience.

But my devotion to the law came at a price. The time I had wanted to devote to the hard questions of life and death, and whether the Christian afterlife really existed gave way to more immediate issues at hand—like whether I could save my client from prison or execution.

As the assistant public defender for the Sixth Judicial Circuit of Florida, I tried somewhere between 140-160 jury trials (I lost count) on cases ranging from petty thefts to murders. But no matter how serious or minor the case, I always felt a heavy responsibility to give each client my complete attention. As a result, I had no time or energy to give serious attention to the issue of whether an afterlife exists.

When I retired in 2016, I needed to come full circle. I needed to go back in time, in my mind, and reconnect with my childhood search for answers to the "afterlife question." This book is a testament to the answers I found—and the questions those answers raise for us all.

PART ONE:

THE THEORY

1. THE CROSSROADS

We stand at a crossroads. Our crossroads cannot be found on any map. It is a waypoint, not of space and place but of mind and time. Modern technological advances have brought humanity to a critical nexus where the gravitational pull of our ideological past, faith, and religious belief are racing headlong on a collision course with humanity's rush toward its future and scientific discovery. Nothing epitomizes this conflict between past and future, faith and fact, religious belief and scientific objectivity more than the issue of whether there is a life after death.

Is there life after death?

We all want to know the answer to that question. There may not be many questions each of us can agree on that matter in our lives, but we are all focused upon that one. That is the question this book will attempt to answer.

Priests, clergy, rabbis, and clerics have told our ancestors that there can be no life after death without religious faith. But let us ignore them for once. I am not talking about the afterlife we have been told exists by our world's many religions. What I am talking about, and what this book is about, is the question:

Does an afterlife exist independent of religious faith?

Is a nonreligious afterlife even possible? If it is, what are the processes by which such an afterlife occurs? None of the world's great religions attempt to answer this question scientifically. Their explanations of an afterlife remain frozen in time—written thousands of years ago in the Bible, the Torah, the Qur'an and other ancient texts.

Most of the world's religions profess the existence of an afterlife but none have ever been able or willing to describe the physical processes involved. Perhaps this is because religion is based upon subjective belief—not objective, verifiable, and reproducible truths required of true scientific inquiry.

Instead of scientific truth, our religious teachings enjoy the influence that thousands of years of indoctrination and ritual bestow upon it. But these teachings are not science; they are but faith. While faith has its power, without science there can be no certainty of truth. Without truth, a religion is mere illusion—a carnival with priests as mere clever showmen and humanity as a congregation of dumb puppets dancing, pulled by invisible strings. Without truth, faith evaporates in the air like moisture in the heat of the sun.

Over the course of the last 100 years, science has shown us what our religious faiths could not: increasingly more revealing pictures of the universe and our place within it. The Hubble Telescope and other satellites give us pictures of the cosmos almost to the very dawn of our universe. Geneticists have mapped both human and animal DNA and RNA. Researchers now program our DNA to resist disease and birth defects. Engineers now program machines (even automobiles) with artificial intelligence. Physicists use massive particle accelerators, such as the one at CERN, to discover ever smaller and more fundamental elements like the Higgs boson and bring us ever closer to replicating the Big Bang.

If we are capable of all this, is not our technology now advanced enough to begin the process of unmasking the greatest mystery of all—the afterlife? If we can spend billions of dollars searching for intelligent life on other planets, we can begin to look scientifically at whether there is life after death.

Does an afterlife exist, or not? Surely, our technologically advanced civilization is mature enough to put the afterlife to the test. But can our science now show us the truth of what lies beyond "the pale frontier" we call death? Is it not time to begin the study of life beyond the grave?

More to the point, are we now ready to take the bold next step in our intellectual and spiritual enlightenment, and face the challenge the following question poses to each of us?

Humanity is ready for this journey. In fact, humanity is desperate for an opportunity to free itself from the intellectual chains of thousand-year-old ideologies that have less to do with individual spiritual development than the manipulation of societies through mind control of the masses.

So, let us turn the question inside out. Instead of whether an afterlife exists, let us pose the following question instead:

Does an afterlife exist dependent solely upon natural laws?

This book presents a biologically based theoretical model of life after death independent of religious faith. This paper explores the possibility that an afterlife does exist—not as a religious expression—but as a purely biological one; dependent solely upon natural laws.

We do not present proven fact. Presented here is still only a theoretical model. However, this theory is supported by modern scientific research. Each of the biological processes necessary for our theoretical model to work is validated by published scientific research, each cited herein for easy reference.

What this means is that the biological processes necessary for an afterlife to exist are present in the human body. Until now, they have remained hidden, like pieces of a puzzle waiting for someone to put them together. The answer to this puzzle begins with a simple question:

What is the objective purpose of life?

We understand there are many religious and philosophical perspectives regarding this question. However, we do not concern ourselves with either religion or philosophy here. Here, those issues no longer matter. Here, we are going back to square one, before all the religious hyperbole and philosophical narcissism. Here, there is only one issue—life itself.

From the very moment life on Earth began as single-celled organisms with cilia reaching out into the external environment for stimuli, existence has been one with the quest for knowledge.

Billions of years later, cilia have become nerve endings reaching out in skin-covered limbs, and stimuli is no longer other bacteria but the universe itself. Nevertheless, the concepts and precepts upon which life began remains with us today.

We, the human ruling class of the modern world, cushion our daily existence with the superficial luxuries of technology. We stand upon the summit of the pyramid of biological evolution. Strip away our science, remove our engineering, and we share the same priority that dominates the existence of every other species on the planet. We wish to survive. We need to survive. Survival is the driving force that propels all we do. Its premise is inlaid within every motivation and every action we take.

Survival for us, like for every species that came before, depends upon one thing—information. We worship information. We spend our lives in the pursuit, collection, preservation, and sharing of information. All human life—from the very sensory inputs of our five senses to the love of history, literature, science, the arts, construction, sports, music, religion, philosophy, law, and politics—is ultimately intimately intertwined with information. The context may change, but the urge to know remains fixed and unwavering.

Why?

Information is nothing more than stimuli evolution has programmed us to rely upon for our individual survival and experiential growth. The information we receive during a lifetime is stored in our neurons as memories. Our brain is a vast memory bank of everything we have experienced—not only in this life but from past lives as well. What is the ultimate application of all this information? What is being created or conserved?

We label this creation "individuality," which every human possesses as a construct of one's consciousness and memories. Every life is unique. Every life has unique episodic sequences of events. These event sequences are stored in our neurons as memories. Every time we consider our past, our memories flow like water over the emotional waterfall of our limbic system. This waterfall of memories through the limbic system constructs our identities, our individualities. This information is shaped by our own uniquely evolving personalities and then formatted as neural and synaptic memories which are in chemical communication with our DNA and RNA codes by evolutionary processes.

Does evolution find individuality of information the main point of human cerebral growth? If so, and if individuality is so important, evolution would need to protect its existence. What biochemical processes would evolution likely create to protect its most important creation—individuality?

This book is written in two parts. Part One presents the theoretical model of "PhiAlpha." This model consists of a complete, multisystemic, biological process which preserves our personal, unique individualities from one lifetime to another. The reader can relax about how difficult this book is to understand. It is very easy. It was written by and for the nonscientist. PhiAlpha applies to everyone, so it is important that everyone can easily grasp and digest its processes—nonscientists as well as scientists.

Part One walks the reader through the scientific concepts of PhiAlpha in layman's terms as much as possible. If the reader is able to understand the concepts presented in this chapter, the rest of the book should be just as easy to read and understand.

Part Two is not scientific at all. Part Two describes how PhiAlpha affected my life. It also presents a commentary on whether PhiAlpha threatens our religious and spiritual beliefs. Will it change the world? Or, has it already done that?

This is a science book written for the average person. It is also a New Age book that explores the edges of our spirituality and religion in this uber-modern age.

Does an afterlife exist—or not? The world's religions have had thousands of years to prove its existence and shown us nothing of objective, reliable consequence. Can science do any better?

2. FAKE NEWS

Even as Nordic Vikings routinely sailed between Newfoundland, Nova Scotia, and other areas of the New World, Christian Europeans were still being brainwashed by their priests and monarchs that the world was flat. The Christian church was threatened by the existence of any New World that might challenge its authority and control over the descendants of the Greco-Roman empire.

It took a leap of faith by Christopher Columbus to put that "fake news" to the test. As a result, Columbus's trek to the New World broke the chains of intellectual imprisonment of an entire civilization and opened a new continent for exploration and discovery.

That was hundreds of years ago, but those same chains still exist today. Those same chains still limit the intellectual freedom of discovery and enlightenment in people all over the world when it comes to thinking about death and what lies beyond.

We die. We die.
Original sin is why.
Religion never lies.
We have always died.
Religion rules the afterlife.
Never ask:
Why?

For thousands of years, this has been the human mantra. Reinforced by countless sermons and rituals imposed upon us by our own religious faiths; this myth has prevented even the most adventurous of human minds from asking the biggest question of all:

Evolution controls our life; does it also control our death?

For thousands of years, our ancestors accepted whatever the religious elites told them to believe about death and whatever follows afterward. Even today, a review of the scientific literature reveals an astonishing truth: a life-after-death science does not yet exist.

But there are questions that logic compels us to answer:

1. Why do we die?

The law of conservation of energy states that energy cannot be lost. Energy can only change its form from one state to another through transitions consistent with natural laws.

So why do we die?

2. What happens to our energy after we die?

If our energy cannot be lost but only transformed from one state to another, what is that state?

3. How is our energy transformed when we die and where does it go? In other words, what are the specific physical processes involved in the transit of energy from our dying bodies to…whatever exists beyond?

These are the questions that any life-after-death science would ask. But, aside from anecdotal reporting from curious emergency room physicians who have witnessed events that have no apparent explanation, scientists have been unwilling to study these issues. As a result, there is no life-after-death science.

The question begs: Why not?

While there have been countless writings on the issue of an afterlife, they have all been religious in nature. The Egyptian Book of the Dead, Hindu Shruthi, Islamic Qu'ran and Christian Bible each have their own explanations of god-hood and the afterlife, but the premise of each is the same. Each writing is a subjective account of a personal relationship with a super-being who vaguely describes an afterlife and the behavior required of anyone seeking to be granted the gift of a life after death.

Only in the leading Hebrew religious texts (Torah and Talmud) do we find an overt reluctance to commit to a set belief structure on an afterlife or *olam ha-ba* (jewishvirtuallibrary.org). It is suggested there that this reluctance was forged out of the Israelites' time in Egypt, when they may have developed an antipathy toward the Egyptian obsession with death.

History teaches us the causal effects of our blind acceptance of theocratic indoctrination. We know what that means. Religious fervor has killed more humans than any other causal factor in history.

The list is as long as it is tragic: ritual human and animal sacrifices dating back thousands of years, Incan and Mayan human sacrifices, the "Christian" crusades, Islamic jihads, the "Christian" Inquisition, the French Wars of Religion, the Holocaust, Ku Klux Klan murders in the American South, and genocide of native North Americans are just a small sampling of the violence humans have inflicted upon their brothers and sisters in the name of religion (and you can see that the predominant offender is the Christian religion, which often is anything but Christ-like).

The beast of violence in the name of religion continues today. Radical interpretations of the Qu'ran spread like cancer throughout the Middle East and the rest of the world, creating a demented ideology of death for misguided fanatics convinced that it is holy to murder innocent men, women, and children.

The beast of religious intolerance and hate spawn death and destruction even today: abortion clinic bombings (Christianity); the Boston Marathon bombing (Islam); the Israeli-Arab conflict is a continuing struggle between religious-based cultures and the nations of the Middle East (Islam and Judaism); and the Fort Hood and Orlando Pulse nightclub shootings (Islam)—to name just a few. No religion is immune; all have sinned in the persecution of others.

Religious indoctrination attacks the mind. A person who was once calm and reasonable before religious training becomes an intolerant fanatic afterward. Religious orthodoxy requires absolute obedience, and that obedience comes at a price—an ever-increasing loss of one's freedom of individual expression.

This loss of freedom of expression is only the beginning, however. Once one's individuality is lost, so too are one's ideals—even basic morality. Eventually, a mob mentality takes over. The result is no longer an individual identity but an inhuman machine willing to kill a complete stranger in the name of religion.

The effect of this religious-created blindness has had its pernicious influence upon scientific expression as well. Over the thousands and thousands of years of human recorded history, we see many expressions of religiously defined interest in the afterlife—but none regarding any nonreligious or scientific alternative beliefs. It is as though when it comes to the issue of life after death, humans can only think in terms of religious explanations. We are blinded by our religious upbringing. It is this very blindness that has kept humanity from looking for an alternative explanation to life after death all these centuries.

The myths and mysteries of life after death have come to us through cave art, fables, clan tales, prophets, priests, martyrs, and scribes since the dawn of history. Human history is, itself, a chronological recitation of our ancestral dependence upon the crutch of religious faith to face each new day. We have all been affected by it. Our perceptions of reality have been so intertwined with religious iconography that we find it difficult to see the universe for what it truly is. In our blind rush toward subjectively perceived emotional safety, we closed our eyes to that which objectively confronts us—the objective universe.

The veil, woven by our ancestors, eventually becomes our own. The deception is planted at infancy by those we trust—our parents. The deception is reinforced throughout life, every season of every year, through community festivals that serve to reinforce the collective faith in the deception—as well as to create cognitive dissonance within any individual whose mind seeks to break free of the deception to search for an independent reality. Community pressure is brought to bear upon any individual who, in confession or to family, admits to being "tempted" to break away.

For most, there soon comes a time when the indoctrination is so overwhelming, the individual pursuit for truth inherent within every soul is forsaken as the price paid for community approval and acceptance. There is then no longer a need for external reinforcement. The deception has taken root within the psyche and one's own mind becomes the principle enforcer of the deception. The objective universe no longer exists. In its place is the fictional universe of theocratic subjectivity.

That is the case for most us. Most of us want, even need, to be liked by our community. Conformity is easier than being cast out as a heretic—or worse—for rejecting the norms of our religious traditions. The psychological pressures to conform a worldview of one's community is often too great to resist for long. Resistance is, in the end, futile.

One's own ideals surrender to the will of one's less curious and more indoctrinated peers.

As independent entities of individual consciousnesses we have a right to pursue our own inner journeys of revelation independent of others. But these rights come with the duty to respect one's own intellectual independence by continuing education of the self and open-mindedness to the opposing views of others.

But religion is not the same as spirituality. We should not confuse the two. Religion is merely how we humans organize ourselves within our communities. Religion is impersonal, public, outward looking, and temporal. On the other hand, spirituality is our personal experience with the universe and our existence within it. Spirituality is personal, private, inward looking, and timeless.

Too often, however, we get the two confused or we sacrifice our spiritual integrity for the comfort (however temporary and superficial) of acquiescing to the social norms inherent in religious beliefs of our own communities. But by so doing, we become complicit in our own phenomenological ignorance. We allow others to tell us what to think and how to express ourselves in actions without any significant resistance. Worse, some have witnessed violence and hate in the name of religious faith without the will to stand opposed. They have become complicit.

Nevertheless, there is hope. Time and events propel us to our shared destiny independent of our finite intentions and expectations. Flickers of light within the gloom of our private intellectual caves have been observed regarding the question of an afterlife. The flickers create shadows. The shadows reveal random activity that begins trending toward a burgeoning realization of inevitable significance.

Thus, we witness the dawn of a new age—the birth of a life-after-death" science.

Toward that end we present what few articles of scientific research on the broad scope of near-death experiences we found pertinent to our purposes. While the papers presented herein cannot be classified as an "afterlife science," they do constitute a representative sampling of what ancillary research exists in related fields:

Peter Craffert, PhD, is a professor in the Department of Biblical and Ancient Studies at the University of South Africa. His 2015 article reviewing the last 40 years of research into nonlocal consciousness argues that research of out-of-body and near-death experience (NDE) have failed to prove that consciousness can exist outside the human brain. Accounts of NDE are not evidence without objective proof. Further, he argues, testimonials of personal perceptions of NDE are anecdotal at best and at worst, hallucinatory.

Craffert's concerns remind us that extraordinary claims require extraordinary evidence, and radical views require radical evidence of the highest quality.[1]

However, a scientist and two medical doctors have reported observations of phenomena that appear to avoid the errors Craffert warns us against and strongly suggest he is mistaken.

Born in Inner Mongolia, Jimo Borjigin received her doctorate at the prestigious Johns Hopkins University in Baltimore, Maryland, and has been an associate professor at the University of Michigan Medical School. Borjigin's work has earned her both the Life Sciences Research Foundation Fellowship and the John Merck Scholars Award.

Studying rats, Borjigin discovered their brains exhibited surges of highly coherent, synchronous gamma oscillations that occurred within the first 30 seconds of cardiac arrest supporting heightened consciousness at death. This high-frequency burst consisted of a striking increase in anterior-posterior connectivity in both theta and alpha wavelengths and exceeded any levels produced during normal, healthy waking states.[2]

Before Lakhmir S. Chawla, MD, was appointed chief medical officer at La Pharmaceutical, he was a practicing anesthesiologist and taught medicine as an associate professor at The George Washington University School of Medicine and Health Sciences in Washington, D.C.

Chawla's research involved human patients at end of life, which he observed in his work as a practicing anesthesiologist at the GWU hospital. Unlike Craffert, Chawla witnessed the process of death firsthand many times. Chawla reports that of the approximately 100 human subjects whose death he witnessed, 80 percent exhibited an unexplained EEG spike at death.[3]

While conceding no one knows what causes these EEG spikes, Chawla concludes this EEG spike observed in so many patients, at death, cannot be a random event and suggests the spike is a telltale signal of a physiological process, the cause of which, remains a mystery.

James H. Lake, MD is a board-certified psychiatrist who has published research in peer-reviewed journals on integrative medicine and psychiatry with emphasis on the role of consciousness and intentionality in human healing. He, too, has observed sudden EEG spikes at near-death events and wonders why they happen.

Lake acknowledges he does not know the cause of, or the reason for, these obviously nonrandom, near-death EEG spikes. To explore this mystery, he proposes what he terms a "testable neural model" based upon his research. He sees the near-death experience as a unique event consisting of the encoding and decoding of information in cortical and subcortical brain regions. He proposes a longitudinal study using sophisticated EEG and brain imaging techniques with advanced data analysis to further explore these EEG surges to determine their origin and purpose.[4]

That is all there is—for now. To put the issue in proper historical perspective, consider this: Regarding life after death, we are where Sir Isaac Newton stood when he witnessed the apple first fall from the tree. All Newton knew was what he observed. Without a theoretical model to go by, he had no hope of understanding the force causing the apple to fall.

That is where we are now. We stand in Newton's shoes, watching flesh die without a clue what force or forces control it, why it happens, or what happens to the energies released afterward. We observe the superficial processes of physiological death. But we have no idea how those processes relate to the natural laws governing the physical processes of the universe.

Today, however, we stand at the doorway of such a discovery. Behind us are all the myths, fables, and mystical hocus-pocus with which we blinded ourselves over the centuries of intellectual ignorance and fear. Waiting for us ahead is the mystery of life in a causal universe that science is only now beginning to explore. Who can say what we will find?

3. PHIALPHA: THE SEED OF LIFE

Simple observation teaches us that life is a phenomenon dependent upon the interaction of multiple processes occurring both within and outside the body. If life after death exists, it too must be dependent upon the interaction of separate processes from multiple biological functions for it to occur. In other words, if we live in a physical universe ruled by physical natural laws, then anything existing within this universe must obey those laws—even the afterlife.

The phenomenon of life after death has become so identified with the nonscientific and the supernatural that to objectively study the process or processes involved in the death, we must distinguish them from the phenomenon itself. Without some means of differentiating the two, we lose focus upon both. This lack of focus and differentiation interferes with scientific analysis by blurring the edges between both until we begin to confuse one for the other.

This problem can easily be avoided by giving the biological process we are describing a name. Such a name will need to respect the scientific nature of its existence and properly symbolize its theoretical function and purpose.

Therefore, the biological process that we believe create life after death shall be termed "PhiAlpha" (ΦA). We use the Greek alphabet because of its traditional use in science. We use "Phi" and "Alpha" because of their symbolic meanings within the Greek alphabet.

The Greek letter phi (Φ) is a combination of two other Greek letters: iota (I) and omicron (O). The circle of omicron symbolizes how our actions in life define the space and meaning of our existence. Who we are as individuals is defined or encircled by the circumference of our thoughts and deeds. The greater the impact of one's life, the farther out from the center of our being extends the radius of the circle. The straight line of iota symbolizes the union of two natural laws:

Natural Law

(1) **The conservation of energy—the total energy of an isolated system remains constant.**

(2) **The second law of thermodynamics—if the physical processes are irreversible, the combined entropy of the system and the environment must always increase.**

We end our symbol of PhiAlpha with the letter alpha (A) because the result of the unified processes involved in our model create a new physical life for the individual. Therefore, when we wish to identify the theoretical model of life after death, we shall use the term "PhiAlpha."

What is PhiAlpha?

PhiAlpha is an evolutionary process of conserving individuality from one life to another. By using multiple physiological and biochemical processes that conform to fundamental laws of nature, individuality is transferred from the dying into the living. It is a harmonic transfer of information obeying classical and quantum mechanical principles.

We define existence itself as a state of being of coherent energy whose main purpose appears to be the pursuit of ever more complex expressions of individuality. We exist eternally as coherent energy, but that coherent energy experiences transformations in the form of changes of phased states of matter (energy expression) that we have come to know as lifetimes.

Therefore, we define life as an intermediate state of existence phased in biological form. One of the purposes of life is the acquisition, reflection upon, and sharing of ever more complex information that enhances the expression of ever more complex forms of individuality.

The origin of PhiAlpha

1. The universe is a relatively closed system of energy and matter. This is the living environment

   existing in a state of increasing entropy (uncertainty).

2. The living environment creates information in the form of sensory stimuli.

3. Into this experimental tableau are placed living creatures. These creatures must use the sensory organs they are provided with to navigate the perils of their living environment. The purpose is simple—survival.

4. Survival of the fittest (natural selection) becomes an end unto itself—evolution. Some believe evolution has no purpose other than survival. They are wrong. Species survival is not the purpose of evolution. Survival is a simple catalyst. Survival is the natural catalyst to force individual living beings to make choices. From one-celled bacteria to humans, all living things throughout time have been forced to make choices necessary to survive. Some succeed and live to pass on their DNA to future generations. Others make wrong choices. They fail to survive.

Sounds like the algorithm for a new role-playing video game doesn't it? But it isn't.

It is real life.

All the choices individual creatures make are necessary for natural selection to function. Choose correctly, and you survive to procreate and improve the gene pool. Choose incorrectly, and you die. Your genes are lost forever.

One cannot study how life evolved from microscopic amoeba to the highly complex human beings we have become without sensing and appreciating evolution's logical progression. There is a definite purpose to evolution. What purpose—beyond ever-increasing complexity of individual expression—we have yet to understand.

Some academicians argue evolution has no purpose beyond survivability. This is akin to saying the universe has no purpose. They lack the vision to see beyond their narrow focus of scientific expertise. They fail to realize evolution is greater than the sum of its parts. They are so flush with excitement over what they have learned that they are blind to the reality that what they know is comparatively insignificant to what remains hidden.

We need more scientists with the humility to appreciate the unknown and less who waste their careers congratulating themselves on what comparatively little they have discovered to date.

The bottom line is that for all we think we know about evolution, it remains an enigma.

We know what evolution has done to us so far; but we have no clue what it will be doing to us in the future. No physical process exists without a purpose. We know nothing about evolution's origin. Where did it come from? How was it created: organically or artificially? How did it become part of our biological, cellular makeup?

Until we can answer those questions, we cannot say we have a clue on the more pressing question: Where is evolution leading us—and why?

5. Evolution creates ever-increasing complexity of information and the biological processes necessary to adapt to it for survival. As the environment increases in complexity, so do our cognitive abilities. One begins to see both environment and evolution as the opposing ends of a rubber band. When one pulls, the other follows. Among the many adaptations that evolution adds to our cognitive abilities, the most important becomes that of memory.

6. Over time, the information stored in human memory becomes the framework within which the consequences of the choices we make become our emotions. This emotional memory is a unique blend of information plus emotion—our individuality.

7. Natural laws prevent the destruction of energy and information.

8. Human death interferes with information conservation (survival of individuality).

9. As evolution must obey natural laws, biological adaptations would be constructed through evolution to ensure that individuality is conserved as unique information.

10. PhiAlpha is what we call the biological adaptations evolution would most likely choose.

The following is an absolute truth:

"The laws of thermodynamics are special laws that sit above the ordinary laws of nature as laws about laws or laws upon which the others depend. It can be successfully shown that without the first and second laws (of thermodynamics) which express symmetric properties of the world, there could be no other laws at all."[5]

Thermodynamics began as a study of how steam behaves under differential pressures and temperatures over time inside confined spaces such as steam engines. However, scientists discovered that the principles governing heat and pressure in such confined spaces are merely applications of fundamental laws governing any process that cause a change of state of being of the particle or particles involved. There are four separate laws of thermodynamics. We concern ourselves with only the first two.[6]

The first law of thermodynamics

"All real-world processes involve transformations of energy, and the total amount of energy is always conserved (stays the same). The universe is a continuum of energy which remains the same regardless of transformation. As to energy, time is symmetrical. The same energy exists for all time, simply changing states over time."[7]

The second law of thermodynamics

"In all natural processes, the total entropy of a system plus its environment cannot decrease. It can remain constant for a reversible process but must always increase for an irreversible one. The universe acts spontaneously to minimize potential outcomes of any process. Therefore, the universe always seeks to maximize disorder (entropy). As to entropy, time is asymmetrical. Therefore, the Second Law of Termodynamics creates a measure of the transformation of energy over time."[8]

The concepts of entropy and information are often misunderstood. Entropy is a term used to describe the relative disorder existing in a closed system. Some prefer to define "entropy" as "disorder." However, we prefer to use the term "uncertainty" instead.

Why?

The answer to this question is critical to our inquiry. The term "disorder" is relatively vague. The term "disorder" can mean something severe, like "chaos" or something minor, like "messy." Neither terms—"chaos" nor "messy"—have the necessary objectivity science requires. Neither can they be objectively measured or analyzed. They are both too subjective to have scientific consequence.

"Uncertainty," however, implies a lack of something objective—information. Information can be measured, analyzed, and transformed from one state to another. Information is the difference between uncertainty and certainty. Information is the balance between uncertainty and knowledge.

To illustrate, consider this thought experiment:

You are walking alone in the desert for hours. You walk until you cannot walk any longer. Stopping to rest and refresh yourself, you look around. As far as your eyes can see, there is only the sunbaked sand of the desert.

On a whim, you reach down and pick up a handful of sand. It is hot to the touch, but at least you know where the sand is. The uncertainty or entropy is low. You know where it is. It is in your hand. You feel it. You know it is hot, dry, and hard. You possess all of the information you need to know what the sand is.

Now, you suddenly throw the fistful of sand into the air. The second the sand leaves your hand, the uncertainty or entropy of the sand increases 100 percent. You no longer possess it. Each grain now possesses its own trajectory, independent of the others. You do not know how long each grain will be in the air or where each will land. This uncertainty increases over time the longer each grain flies in the air.

But here, as everywhere else in nature, there is balance. While uncertainty increases over time, so too does information. So, you watch and observe the grains of sand as they arc their way above the desert floor until they land. Your sense of sight gives you information on where and how far each granule of sand travels and where each one lands. While uncertainty gradually increases as each granule travels through the air, this uncertainty is continuously being resolved through the information from your sense of sight. A balance is struck. What entropy would conceal, information both reveals and conserves.

Our universe, being a closed system, is subject to the laws of thermodynamics like any other. At the time of the Big Bang, entropy (uncertainty) was relatively low. Over time, however, the level of entropy has increased. This change in the degree of entropy can be measured. Information, over time, creates measurements of change to our state of being. This proposition leads us to the surprising but logically inevitable conclusion that all knowledge is reducible to classical mechanical analysis.

Since the second law of thermodynamics provides a measurement of change over time, scientists have concluded that this second law governs the conservation of information. Information is the result of any observation or measurement. One may even conclude information equals existence and vice versa.

When any object undergoes a transformation, it creates information about its change of state during its transformation. All objects, including biological lifeforms (humans) create information describing the relative state of being during their existence. Information is created whenever any object undergoes transformation within a system as it is being transformed.

In 2013, a research team from the People's Republic of China, led by Dr. Baocheng Zhang and Qing-yu Cai, PhD, published a paper on information conservation that is a beautifully articulated presentation of the principles involved. Their conclusion speaks not only to classical mechanics, but to the realm of quantum mechanics as well. Both Zhang and Cai are professors of physics at Hong Kong Polytechnic University. Their report states:

"Information is physical, it cannot simply disappear in **any** [emphasis added] physical process. This basic principle of information science constitutes one of the most important elements for the very formulation of our daily life and our understanding of the universe." [9]

The reader should not ignore the word "any" in the above quotation's first sentence. Zhang and his team are trying to explain the broad implications of information conservation theory.

When they state information is conserved in "any" physical process, they mean just that—any. That includes the physiological lifecycle of humans.

The human physiological lifecycle is a physical process just like any other, and so it must conform to the law of information conservation. The information within the universe cannot be lost or destroyed. Theoretically, if information could be lost or destroyed, then the Big Bang itself could be undone. That is an impossible paradox.

The father-son team of Ryszard Horodecki, PhD, and Michal Horodecki, PhD, are just two of an entire family of physicists from the University of Gdansk, Poland. Michal is internationally recognized for his discovery of quantum-state merging and for co-discovery of the Peres-Horodecki criterion for testing whether two quantum states have become entangled. Together, the two Horodeckis reformulated the traditional second law of thermodynamics to better conform to quantum theory. Their redefinition of the second law of thermodynamics according to quantum theory is as follows:

"In quantum mechanics, one cannot either delete or clone [quantum information]. The Second Law holds: entropy cannot be decreased in a closed system. Again, the Second Law can be stated

in a stronger way: that eigenvalues of density matrix do not change under quantum evolution. And this is actually the information conservation principle."[10]

From this redefinition of the second law came two principles of information conservation:

(1) The No-Deleting Principle

"In a closed system, one cannot destroy quantum information. Quantum information can only be moved from one sub-space location to another. In a closed system, even orthogonal [opposite] states cannot be deleted alone, given unitary evolution. Once entropy or information has increased, it cannot be decreased—it can only be transferred to a separate location."[11]

(2) The No-Cloning Principle

"Quantum information cannot be 'leaked-out' while keeping the original information intact. This is true for both open and closed systems, as well. This principle is already provable using unitary evolution dynamics. Once information is increased, it cannot return to its original state."[12] In other words, quantum information cannot be duplicated. Therefore, information is unique unto itself.

A physicist, Marco Roncoglio's field of expertise is quantum mechanics and theoretical physics. He is a past visiting fellow at the University of Munich's Department of Physics. His paper, "On the Conservation of Information in Quantum Physics," introduced the concept of coherent information being conserved in a quantum state. According to Roncoglio:

"...[P]ure states contain more coherent information than mixed states, as the missing information has been converted into correlations with the environment. We will find that the coherent information, associated to genuinely quantum phenomena, is indeed conserved under unitary processes."[13]

Roncoglia's "correlations" sound very much like the neurally encoded memories we observe in the human brain. Put simply, what we experience as neutrally encoded stimuli is Roncoglia's "pure information."

This information is "pure" because it is raw sensory input. There is no emotional bias affecting its coherence. Through contemplation, analysis, or simply impulse, we mix this "pure" sensory input with the emotional resonance created from memories from past experiences. In this process, some of the pure data is "lost" to our consciousness but never really lost. While lost to our conscious awareness, the original, sensory information remains conserved under quantum processes.

That is how information conservation becomes important to our human existence. It places the information we use to create our own unique information—individuality—within the framework of quantum mechanics and the second law of thermodynamics.

PhiAlpha is based upon two fundamental principles:

(1) Evolutionary adaptations of the human brain created the process of creating unique information we call "individuality." This individuality is unique information created through the interaction of new stimuli (pure information) with emotionally loaded memories of past experiences (mixed information) and;

(2) The quantum mechanical application of the second law of thermodynamics upon this unique information we call "individuality" compelled the evolutionary adaptations that created PhiAlpha—the biological processes that conserve individuality from one life to another.

Objections to PhiAlpha

There are three basic objections to the PhiAlpha theoretical model. The first objection is that PhiAlpha is not testable and that until PhiAlpha is testable, it must be regarded as nonscience. People who believe this are mistaken. One aspect of the PhiAlpha model has not only been tested, it has already been proven. It is as follows:

Before her tragic death on her 27th birthday in 2009, Jharana Rani Samal, was one of India's brightest young physicists. She and her research team conducted an experimental test of the information conservation's no-deleting theorem (her team called it the "no-hiding" theorem) by using nuclear magnetic resonance and quantum state randomization of a qubit to show that the missing information can be fully recovered. This test proved that if any physical process results in the apparent loss of information, the missing information is not lost but merely relocated.[14]

In her honor, co-researchers Arun Pati and Anil Kumar had her work published posthumously in 2011. Pati has been professor of quantum information at one of India's leading physics research labs. In 2008, he co-edited *Quantum Aspects of Life*, which included the singular honor of having its forward written by none other than England's world-renown physicist-mathematician Sir Roger Penrose. Pati's articles have earned publication in such noted science journals as *Nature* and *Science*, and he is an elected member of India's National Academy of Sciences.

True, none of these researchers understood the significance of their discovery at the time. But no one has ever before considered the application of information conservation principles upon the human memories that create our unique identities.

The second objection or challenge to PhiAlpha is based upon a perceived lack of evidence. Some claim PhiAlpha does not exist because no evidence of its existence is observed. There is a false logic at work in this supposition, however. For example, there is no evidence that aliens exist on other planets. Does that lack of evidence convince us not to search? No, of course not.

We do not need tangible physical evidence of the presence of alien life on other planets because we believe they do exist. The weight of the logical inferences we have drawn from our life experience compels our faith in such a conclusion. Such is the case with a biological process for life after death. The weight of logical inference based upon our life experience forces the conclusion upon us.

The third challenge arises out of the problem of transferring information from a dying body into one that is living, one that is about to be born. Without a transport mechanism in place to secure our memories and transfer them from our dying body to an unborn one, this is all just a waste of time—isn't it?

This is true.

Furthermore, wouldn't this transport mechanism have to be a quantum mechanical one at that?

Yes, this is also true.

Then are we not wasting our time? It's impossible to find such a mechanism, isn't it?

Wrong.

The transport mechanism we need already exists. The transport mechanism we are looking for is found within every cell of our bodies—biophotons.

The team of Siyuan Shi, Prem Kumar, PhD, and K.F. Lee proved the biophotonic transmission of information (energy) in 2017. Kumar has served as professor of physics at Northwestern University's Center for Photonic Communication and Computing. Together, they generated a polarization-entangled two-photon state in an ensemble of fluorophones. According to their report:

"Recent development of spectroscopic techniques based on quantum states of light can precipitate many breakthroughs in observing and controlling light-matter interactions in biological materials on a fundamental quantum level."[15]

Biophoton packets can be entangled to transmit coherent information almost instantaneously and, seemingly, without limit. Researcher Tristan Tentrup, a promising Ph.D. candidate from the University of Twente in The Netherlands, and his team claim that "encoding information in the position of single photons has no known limits. ..."[16]

Even our military has recognized the importance of using photons as a carrier of information.

In a rare unclassified research report, the United States Defense Advanced Research Projects Agency (DARPA) describes the photon as "... a fundamental carrier of information, possessing numerous information carrying degrees of freedom including, frequency, phase, arrival time, orbital and angular momentum, linear momentum, entanglement, etc."[17]

The purpose of the DARPA program (dubbed InPho) is to pursue the fundamental science and applications of photonic communications (biophotonic and non-biophotonic) in communications and imagery transfer.[18]

DARPA's application of photonic communication for military communications and imagery provides a near-perfect match with our PhiaAlpha model of neural-memory-information processing within our own human brain at death. Later in this book, we apply this knowledge of biophotonic transportation of information to create our theoretical model of life after death.

We conclude that the process humans have termed "death" is not an end of life but a transformation of the coherent energy created by biological life when the physical body decays. Without PhiAlpha, the energy of the biological lifeform would lose coherence at death, causing the dissipation of consciousness along with all the knowledge and memories collected during life. PhiAlpha creates a lifeboat of energy coherence for memories to remain coherent between the end of biological life of the old host and the fetal beginning of the new.

PhiAlpha and Consciousness

Does this mean consciousness is being transferred from one physical body to another?

No.

PhiAlpha is separate from consciousness. The study of PhiAlpha may or may not yield comparative information as to consciousness, but its purpose is distinct from that of consciousness. PhiAlpha is a process intended to conserve individuality as an expression of ever more complex information—which may be an intended or unintended consequence of consciousness. But they are not the same.

Consciousness, we believe, requires self-awareness—which is more than simple identity. Consciousness appears to be different than PhiAlpha, which is merely an individual identity matrix built upon biochemical constructs called memories and the mechanisms evolution created to preserve their existence.

4. EVOLUTION: THE QUIET CREATOR

There is but one interface between life as we know it and the laws of conservation of energy and information. We call this interface genetics or evolution. While DNA and RNA control and predetermine form and context for all life on Earth, these genetic codes must obey the same fundamental laws of the universe as any other physical process.

Though we do not yet know how evolution began or what (or who) created it, we do know evolution is a process created from within the fundamental laws of the universe. Therefore, its function and biology must act in accord with those universal laws. We are only now beginning to understand genetics. The relationship between genetic evolution and the forces and elements that combine to create life remain unknown.

However, it would be logically inconsistent to conclude that the process of biological adaptation to preserve life (evolution) is separate and apart from whatever forces in the universe spawned life in the first place. The two must be intertwined. They must exist concurrently and interact in balance with one another like the pull-and-give of two sides of a rubber band or string.

Evolution is not fully appreciated yet as the universal force of nature it may be. To not include evolution with the other major forces of nature (electromagnetism, strong and weak nuclear forces, and gravity) ignores evolution's impact upon all life. Life itself is a force of nature, interconnected with each of the other major forces in ways we do not yet fathom.

Our theory is that PhiAlpha is an evanescent, emergent genetic trait evolution constructed over time, applying fundamental processes of thermodynamics, physics, and biochemistry to comply with the fundamental conservation laws of energy and information.

The physical universe is a flux of matter and energy that constantly changes states over time. Therefore, the state of existence we humans know as our present physical life is not meant to exist forever in the same form.

By the same laws, life cannot cease to exist either. That would not be logical. Logic tells us that existence continues in some form or state of matter and energy after physical death. The laws of energy and information conservation mandate that information continues to exist—in one form or another—forever.

PhiAlpha is the bridge evolution created to comply with these two laws. Human evolution resulted in individuality as the genetic trait best suited to organize and preserve information under natural law. PhiAlpha is the logical evolutionary adaptation of biological processes needed to form a coherent phased state of energy capable of receiving, prioritizing, storing, and transferring information (memories) in sufficient quantity and quality to isolate and preserve individuality between human physical lifetimes.

Once the physical body dies, incorporeality diffuses the information stored in the brain like steam escaping a boiling pot of water. To conserve information beyond death, evolution needed a method of creating a coherent pulse of information capable of surviving the loss of corporeality until reorganizing itself in a living host.

PhiAlpha must, therefore, be a combination of genes, proteins, and amino acids whose expression need not occur during normal life at all. They may lie dormant during normal life and only become expressed as the body encounters a life-threatening event.

As the body reacts to the threat, a biochemical "trigger" would cause the expression of genes necessary to organize, prioritize, store, and transfer information (memories) from the brain's neural net into a coherent energy state lasting long enough preserve the information as it transits from the dying host and then into a living fetus.

Evolution's capacity to upregulate genes at near death or even postmortem was documented in 2016 by Peter Noble, PhD, a professor at the University of Washington, whose expertise is in the field of gene expression, molecular genetics and DNA sequencing.

In one experiment, Noble studied how zebrafish and mice genes behaved after death. To his surprise, neither sets of genes decayed when their fish and mouse hosts died. Instead, they suddenly began functioning only after death. This presents a rather bizarre mystery. Why certain genetic profiles turn on or are upregulated (including thanatotranscriptome) only after death is unknown:

"Upregulation of genes demonstrates formation of new molecules post-mortem. Sufficient energy within cells of dead animals exits to maintain gene transcription for up to 96 hours after death. An unknown regulatory network appears to be still turning "on" and "off" genes in organismal death..."[19]

Noble's team also found odd the timing of gene upregulation at death. While most of the genes "turning on" at death did so immediately at death, some did not turn "on" for 24 to 48 hours post-mortem:

"These differences in timing and abundances suggest some sort of global network is still operating. ...What makes gene expression of life different from gene expression at death is that post-mortem upregulation of genes offers no obvious benefit to the (now dead) organism. We argue that self-organizing processes driven by thermodynamics are responsible for the post-mortem upregulation of genes."[20]

Noble and his colleagues correctly point out the mystery regarding the upregulation of genes post-mortem without any apparent benefit to the now deceased animal. Evolution is never without reason, however.

Certainly, the ability to turn specific genes "on" at and after death is a necessary component for PhiAlpha to operate. That this capability is found in lower life-forms is evidence of evolution's interest in the process of selective expression of genes at and after death. We see evolution at work, preparing individual processes in lower life-forms to be brought together in higher life-forms.

Evolution is a zero-sum process. Natural selection is all or nothing. If the genetic code of a species maximizes its group fitness for survival (the dinosaurs, for example), it will be able to pass those genes along to its children indefinitely (unforeseen asteroids the exception, of course). If, however, a species lacks group fitness for survival, ultimately there are no children. This species becomes extinct.

Group fitness for survival has been evolution's mantra—until the arrival of humans. Humanity's ascension over the gene-dome as Earth's dominant species caused a fundamental change in evolution's priorities.

Evolution itself is evolving. While continuing to apply its original code of group-fitness, evolution has slowly but gradually built something uniquely human—individuality. Gerald H. Jacobs, PhD, professor of psychology at the University of California at Santa Barbara explains:

"Humanity is in the process of evolving from collective uniformity to increasing variation and diversity. This movement has gained impetus from the growing recognition that the overall strength and sustainability of the collective is proportionate to the value it accords to each individual human being and the active support it lends for full development of each individual's unique, creative potentials..."[21]

Human individuality did not appear out of thin air or overnight. Evolution has been manipulating the trait of individuality for millions, if not billions, of years. True, evolution's methodology may have begun as a shotgun approach toward biological adaptation, but its history reveals its singular purpose to be something quite different.

The major transitions theory of evolution argues that evolution is an ongoing process driven by a small number of major evolutionary transitions. In each transition, a small number of individuals that were previously able to survive and reproduce independently from the group create a new and more complex lifeform. Genes, for example, cooperate to form genomes. Archea and eubacteria formed eukaryotic cells. Cells cooperated to form multicellular organisms.[22]

Evolution has been using individual mutations and group-fitness as an intertwined pair of global trait constructors. One counterbalances the other, but both are essential to the overall process. In 2015, Stuart A. West, PhD, Roberta M. Fisher, PhD, and their research team identified two conditions defining an evolutionary transition in individuality.

Stuart has been professor of evolutionary biology at England's prestigious Oxford University. Fisher is associated with the Department of Ecological Science at Vrije University in Amsterdam, The Netherlands.

First, entities capable of independent replication before the transition can replicate only as part of a larger unit after the transition.

This is called by several terms: mutual dependence, interdependence, contingent irreversibility.

Second, a relative lack of within-group conflict occurs such that the larger group can be thought of as a fitness-maximizing individual (organism). An example of this condition would include a group of cells that make up a multicellular animal as a single organism. When these two conditions are met, West's team hypothesizes, evolution can produce a higher-level individual (organism).[23]

While no species has been individualized more than humans, the trait itself has been in development for untold millions, if not billions, of years. Observation alone confirms that individuality, as a trait, shows increased expression as one moves up the evolutionary ladder toward humanity. The higher and more complex the lifeform the greater the expression of individuality within the species.

As with any new mutation within a species, individuality had to survive as a trait among competing traits, especially the trait of group fitness. In 1998, the noted evolutionary biologist and author of *Darwinian Dynamics*, Richard Michod, PhD, of the University of Arizona, presented an explanation for why individuality prevailed:

"What explains the individuality of organisms? For an organism to be an individual, or evolutionary unit, requires functions at the organism level that protect it from conflict within...natural selection at any level requires heritable variations in fitness at that level. To emerge as an evolutionary individual, cell groups must acquire the means to guarantee fitness heritability at the group organism level."[24]

For evolution to protect its survivability, individuality must have appeared as a very important trait necessary for survival of complex species. To support this theory, Michod identified germline modifiers and self-policing modifiers as the first uniquely organismal functions that evolve during the transition from unicellular to multicellular life. These modifiers protect the evolved (mutated) cell from being attacked from within the pro-unicellular enzymes.[25]

We humans are billions of years ahead of simple multicellular organisms, however. How evolution has modified us over those eons of time remains unclear.

However, Cory McLean and a team of Stanford University researchers have been attempting to identify molecular events particularly likely to produce significant changes in human genetic structure—such as deletion of DNA sequences that were otherwise highly conserved between chimpanzees and other mammals. They were able to identify more than 500 such deletions in humans, which are almost exclusively from what are called "no coding zones." He concludes:

"Humans differ from other animals in many respects of anatomy, physiology and behavior; however, the genotypic basis of most human-specific traits remains unknown."[26]

Evolution and Individuality

We define the term "individuality" to mean the identity matrix created when we subjectively prioritize those memories upon which our self-image depends.

This identity matrix changes over time as we acquire new experiences and emotional responses to those experiences. It would take days to go through all our memories to isolate only those upon which we identify as necessary to describe who we are. PhiAlpha collects, reviews, and prioritizes an individual's memories almost instantly.

(1) Individuality is more important than most other genetic traits such as hair color, eye color, facial features, bipedality, opposable thumbs, and the folds and ridges of the human brain's cortex for greater brain area and greater complexity of thought despite the physical limitations of the human skull. Indeed, individuality is equal with natural selection and group fitness as fundamental algorithms of evolution.

(2) Evolution has been refining ever more complex forms of individuality from the time multicellular organisms emerged from the unicellular. We cannot expect evolution to stop now.

(3) Individuality has proven its survivability over time and must be recognized as a dominant trait. Recorded history confirms individuality's irreversibility as such a trait. Despite repeated efforts over history, no tyrant nor dictator has yet been able to crush the concept of individual freedom from human consciousness. Individual freedom of expression is an irresistible urge within the consciousness of all humans—an urge evolution has gone to great effort over eons to create and to protect.

Evolution has one major problem with individuality, however. The more complex the brain becomes, the more difficult it becomes to pass along individuality through one's genes. Sexual reproduction requires two participants and is, therefore, a group-oriented process. Eventually, individuality no longer remains a group process but begins to be isolated within individuals. The more complex the brain, the more specific the experiential memories become.

The concept of sexual reproduction is a group-oriented process requiring more than one individual to carry on the genetic code of the species. This process creates a need for individuals to interact as part of a social order or group to find mates. Individuality is revolutionary in the sense that it no longer requires group sanction or group interaction to thrive or to survive. Individuality is anti-group evolutionarily as well as socially.

At some point, the development of human individuality conflicted with two of the laws we mentioned earlier: conservation of energy and information. This likely occurred when brain complexity permitted sufficient memory storage to make these memories more valuable to the survivability of the human species than traits passed along genetically from one generation to the next (eye color, physical features, etc.).

How does one pass along a memory from one generation to the next?

Cloning does not work. We can clone the genes that make up one's body and even brain structure, but we cannot clone one's individual life experiences. Even if we could do that, we could not effectively clone an individual personality.

Human individuality cannot be cloned because each person prioritizes memories differently by subjectively chosen criteria. We are the product of the experiences we subjectively choose to motivate our emotions and behavior. One cannot duplicate another's choices by cloning.

Nor can we "download" our memories to the next generation genetically. Personality cannot be transferred to progeny through procreation. Parenting and socialization are the only ways that the identities of the individual and of society can be transferred from one generation to another.

Therefore, if individuality is a necessary trait evolution needs to preserve to comply with the laws of conservation of energy and entropy how does evolution construct such a process as PhiAlpha?

Evolution and the Human Brain

The answer lies in the evolution of the human brain. Evolution has made humans special. We are special because of our brain. Recent research bears this out. The human brain size tripled 1.5 million years ago, with the cerebral cortex expanding at a much faster rate than other areas of the brain.[27]

The human brain is no longer organized the same as the brains of other mammals and primates. This reorganization of central components of the human brain is as important to the brain's development as was its increase in size.[28] It has been suggested that this reorganization occurred due to the human brain's need to manage increasingly complex social and interpersonal interactions and relationships.[29]

In addition, Todd Preuss, of the Yerkes Primate Research Center at Emery University, found more than 100 separate genes differentiated human and chimpanzee brains. He discovered that human brains possess remarkably different spatial patterns of genetic expression compared to both chimp and macaque brains.

Preuss admits he does not yet understand what these differences mean in terms of functional organization, though he does suspect that many of these changes involve upregulation of genes in areas controlling human metabolism, synaptic organization, and synaptic function. These genes are creating a more dynamic brain. As Preuss put it, "The human brain seems to be **running hot** in all sorts of ways [emphasis added]."[30]

If the human brain is indeed "running hot," we do not yet know where it is running to—or why. We do not know evolution's ultimate purpose. Some suggest evolution has no purpose. They are wrong. No biological process exists without purpose. Natural selection makes any other outcome logically impossible. Even if the only purpose of evolution were to be species survival, then that is a purpose.

We do not know evolution's ultimate purpose, but we can speculate. Evolution never operates without a specific plan based upon improved functioning it deems necessary for increased survivability. Research suggests that evolution is directing humanity toward unlimited cognitive and communicative capabilities with the intention of enhancing the expression of individuality within each of us.

Evidence of this is documented in the research of Chet Sherwood, PhD, and his team. Sherwood has been professor of anthropology at The George Washington University in Washington, D.C. After obtaining his doctorate from Columbia University, Sherwood has researched brain evolution in mammals and is affiliated with GWU's Mind-Brain Institute for Neuroscience.[31]

Sherwood identified genetic differentiations between the human brain and those of our primate and mammalian ancestors. The most significant difference between humans and lower species is that of language. He and his team concluded that language is "a unique behavioral adaptation in humans. ...[T]he unique brain growth trajectory of modern humans has made a significant contribution to our species' cognitive and linguistic abilities."[32]

From this research, we can conclude evolution appears to have chosen individuality as a dominant trait in and of itself. Moreover, evolution has chosen the human brain as the most efficient means in which to preserve and refine individuality further. We can express this relationship as a formula:

$$I = (C_1 \ L_1 \ M_1 \ / \ t)$$

Where the letter I stands for the completed trait of individuality; C for cognition; L for language; M for memory and t for time. The subscript 1 below C, L, and M symbolizes a sufficient complexity of these variables at a level sufficient to achieve individuality. Cognition, language, and memory are all evolutionary components of intelligence, and their interaction over time creates individuality.

Why is individuality so important to evolution's plan for humanity? We suggest that the reason is found in the path evolution has taken from primordial times. Evolution is not sentient in the way humans are, with subjective, experiential involvement that mandates varying degrees of bias. Evolution has no subjective identity or intention.

While evolution may use survivability to separate the wheat from the chaff, so to speak, it seems to being using hidden algorithms to change us into something far different from the rest of Earth's species.

One such hidden program, as it may be called, is focused upon ever-increasing awareness or sentience. Natural selection and survivability are just means to an end. They are routers, so to speak, guiding evolution away from species that it would be wasting its time upon and toward a species most suitable to use expanded awareness as a tool and to survive and thrive with this tool.

Individuality is the ultimate application of increased awareness. Evolution has chosen humanity as the only species possessing a brain of sufficient complexity and size to express itself in art, music, song, poetry, literature, religious rituals, sport, and complex forms of interpersonal communication. These traits require unique expressions of individuality.

The greater our awareness of ourselves and our environment, the greater ability we possess to express our individuality. Every emotion that evolution has crafted for humans only serves to enhance and enrich the individual experience of life. Our universe is a symphony of mathematically eloquent logical sequences of ever-increasing complexity.

It is, therefore, logical that evolution would not permit human individuality to be lost. If evolution spent billions of years creating ever more complex means of individual expression, it is logical that evolution would not accept losing its creation to the transience of human lifetimes. There must be a means of isolating that individuality and then transferring it from the body of the dying to that of the living.

How can that happen?

5. INDIVIDUALITY: CHILD OF OUR MEMORIES

To answer this question, we begin with the building blocks of individuality—memories. Memories are experiences created, stored and transmitted through neurons and synapses in our brain. Memories are what the human brain uses to craft the uniqueness of our individual personalities over time. Individuality requires both working memory (day-to-day memory) as well as long-term memory (those memories stored over years or even a lifetime). Working memory allows us to function normally through our daily routines, carry on meaningful relationships, and make rational decisions as we live our day-to-day lives.

Nevertheless, we suspect the memories PhiAlpha prioritizes to preserve individuality are long-term memories. Long-term memories are stored in the cortex, in what is called the gray matter. It is this area of the cortex that appears in photos and diagrams as though folded over upon itself.

Evolution caused the human brain's cortex to fold in upon itself to create more area for greater computational power within the confines of the human skull.

The gray matter is where all the neurons and synapses are located. The folds create ridges called gyri (gyrus, singular) and valleys called sulci (sulcus, singular). Underneath the gray matter is a layer of white matter, which is composed entirely of fibrous axons that act as wires connecting the neurons of one fold to those in another fold.

In healthy humans, these axons are myelinated (covered in an insulating sheath allowing the signal to travel easier and faster).[33] However, at least one study suggests long-term memories are stored in very specific brain cells located in only one area of the human brain—the dentate gyri of the hippocampus.[34]

To fully appreciate the interaction of PhiAlpha and individuality, it is necessary for a brief review of how long-term memories are created and stored. Processing loads within the human brain require an astronomical number of neurons, synapses, and glial cells. The human brain contains over 200 billion neurons with up to 10,000 types of neurons. Each neuron is connected to between 5,000 and

200,000 other neurons. It has been suggested that there are more neural connections in the human brain than there are stars in the universe.[35]

The neurons interact with synapses, axons, and glial cells as signals are transmitted from the brain to and from parts of the body. Synapses carry information by electron and/or biochemical ion transfer between neurons. Neurons and synapses are mostly located in the gray matter of the cortex (the part of the gray matter that is folded creating gyri and sulci). Axons are in the white matter part of the brain lying beneath the gray matter.

We can regard neurons as batteries of stored electrochemical energy. Synapses are like switches, turning on and off as needed to transmit the electrochemical energy into signals. The axons are the wires over which the signals travel. It reminds one of the structure of our modern microprocessors, doesn't it?

It works in the same way. All cells in the body function through electric polarization of molecules. Ion transfer across cell membranes affects voltage levels, which result in either polarization or depolarization of the cell. Memories are contained and moderated in the cells of neurons and synapses. Each neuron may be an individual event memory.

Memory capacity of each neuron is determined by the strength and size of its attached synapse. Synapse size is dependent upon how often each synapse is "fired" by an electrical or biochemical ion signal called an "action potential." Action potentials (more commonly known as nerve impulses) are created by both the voluntary (conscious thought) nervous system or the involuntary (no thought involved) nervous system. Action potentials "fire" when sufficient energy is created within the neuron to send a command or when we relive an experience or consciously think about something. Therefore, synapses increase and decrease in size depending upon how many times the brain uses them.[36]

Researchers estimate every synapse in the brain increases and decreases in size every two to 20 minutes. Total memory storage of the human brain depends upon both the number of neurons and their synapses and the number of distinguishable synaptic strengths. Each synapse is estimated to possess the equivalent of 4.7 bits of computer memory.[37] Multiply 4.7 bits of data per synapse by the number of possible neural-synaptic connections, and the total amount of memory storage space in the human brain is staggering.

All neurons send and receive messages electrochemically by metabolically powered ion pumps that combine with ion channels inside the cell's membrane. These channels generate differentiated electrical charges (interchanging between positive and negative charges of varying strengths) between the interior and exterior of the neurons. Theses electrical charges are created by ions (sodium, potassium, chloride, and calcium atoms, which possess differing levels of electrical charge).[38]

The change between positive and negative ion charges creates a biological alternating current

similar to that found in the common motor. This process creates the energy needed within the cell to pump or move the neural signals from one part of the brain to another.

When an ion crosses the boundary of a neural membrane, the voltage it releases fires an action potential. The voltage required for the firing of an action potential is about -55mV (millivolt). The energy pulse released by the firing of the action potential travels along the cell's axon, activating synapses with other cells (including other neurons).[39]

The electrochemical process that allows the neuron (and all other cells) to transfer electrical signals is called the sodium-potassium pump. The push-pull of osmosis into and out of the cell by this "pump" causes differences in the electrical charge between the interior and exterior of the cell membrane. This push-and-pull effect forces sodium ions within the cell to be repelled outward while potassium ions are propelled inward.

Until recently, it had been assumed that the only function of the sodium-potassium pump was to monitor and control local voltage levels within and around cell membranes. However, the sodium-potassium pump is suspected of being much more than that. Some researchers believe the pump may control cerebellar neurons and may also serve as a computational processor or computer in both the cerebellum and cortex.[40]

Memory recall is a complex electrochemical process. It is this process that creates what is called the brain's bioelectric field. This bioelectric field is created as follows:

When memories are recalled from neurons in the gray matter, their signals travel downward from the gyri of the gray matter into the white matter. Once in the white matter of the brain, axon fibers route the signal over and then upward from the white matter back into the gray matter. Now back in the gray matter part of the brain, the signal is transferred to the appropriate neurons connected to those memories that are stored in other folds within the cortex.

This routing of electrochemical neural signals downward into the white matter and then back upward into another area of gray matter takes the electric field created by the transfer of electrochemical ions in the cortex and spreads it out into the cerebrospinal fluid pulsing over and through the cortex.

Cerebrospinal fluid is an electrically conductive medium, which means it transfers an electric current easily. We believe the down-across-up motion of neural electrical signals through the wavelike folds of the cortex generates an electromagnetic wave through which neural information flows like a current.

This electromagnetic wave interacts with the electrically conductive medium of the cerebrospinal fluid (CSF) to generate and sustain an electromagnetic field capable of supporting a state of phased energy within the brain and spinal cord we term the neural network (neural net). We suspect this neural net serves as an antenna transmitting and receiving biophotons containing individuality in the form of memories stored within the biophotons as coherent bits of information.

How is this accomplished?

We have examined evolution's emphasis upon brain capacity and its relationship with cognition, language, and ultimately expanded awareness leading toward ever-increasing complexity of individuality. We have explored the relationship between individuality and memory, and how memories are constructed. We have demonstrated why individuality cannot be cloned or transferred genetically from one generation to another like eye color or asthma.

If PhiAlpha exists, it does so not as a mystical force independent of physical laws of nature or the universe but as a child of universal laws and processes dependent upon them. PhiAlpha would obey the laws of natural selection and fitness for heritability and survival just as any other process evolution creates. Because PhiAlpha would arise from and out of our own physical bodies, it would obey the same laws of thermodynamics and quantum mechanics as any other physical state would and does.

We cannot expect PhiAlpha to "shout out" its existence at us with one or several specific clues to its existence. PhiAlpha is the forest hiding amid the trees, so to speak.

To find PhiAlpha, we must search for its fingerprints upon the common, ordinary physiological processes we take for granted every day. We need to know what to look for. To do this, we must hypothesize what biological functions PhiAlpha would need to accomplish its task of preserving individuality as memory information. Then we must look within the human body to see if there is any evidence of such processes. If those processes exist, we can deduce PhiAlpha's existence, even if we are presently unable to observe its behavior.

What is the purpose of PhiAlpha?

We understand that evolution employs group fitness and natural selection with the individuality of single mutations to create a counterbalance of spiraling complexity in organisms. We also understand that evolution has selected *Homo sapiens* as a species to be separated from the herd of other species for special attention. This special attention includes maximizing complexity of cognition, language, and memory—creating ever more complex expressions of individuality.

Individuality has not only become the most important trait evolution has created for humans, human history has shown that individuality has become an end unto itself. Therefore, the purpose of PhiAlpha must be the continuation of coherent individuality stored within neural-encoded memories.

PhiAlpha is more easily understood by differentiating its component processes. PhiAlpha components include: (1) **PhiAlpha Activation**; (2) **Memory Organization**; (3) **Memory Prioritization**; (4) **Memory Encoding**; (5) **Memory Emission** and (6) **Memory Reception**.

We will examine each in order.

6. PHIALPHA ACTIVATION

PhiAlpha cannot be "on" all the time, otherwise its system requirements would dominate other physiological systems prohibiting normal life function. PhiAlpha must, therefore, lie dormant during normal life events and require a specific genetic-upregulating stimulus or event for expression.

More simply put, PhiAlpha cannot become active every time we catch a cold or get the flu. If that were the case, our bodies would not be able to function. Even abnormal events such as a broken bone or disease will not initiate PhiAlpha. While serious, each of these conditions are survivable with proper medical care.

Rather, PhiAlpha activation must require an extraordinary life-threatening event. Further, PhiAlpha cannot be a simple one-step or even a two-step process. If this were the case, our minor illnesses and routine accidents would cause PhiAlpha to rev up too easily and too soon. The added physical stress of PhiAlpha's start-up would only exacerbate otherwise routine illnesses and injuries.

To guard against premature PhiAlpha start-up, evolution would need to create a specific set of conditions necessary to trigger PhiAlpha as well as a biological mechanism to control its performance. Therefore, PhiAlpha would need to distinguish between a life-threatening but reversible crisis and the beginning of the death process itself.

Evolution has created the perfect mechanism to serve as a PhiAlpha trigger and monitor—the brain's limbic system:

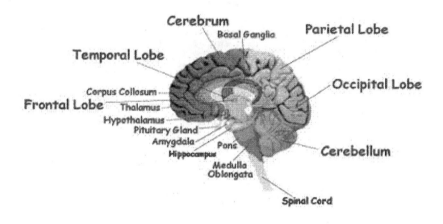

Illustration No. 1

Illustration No. 1 reveals the limbic system (hippocampus, thalamus, amygdala, hypothalamus, basal ganglia, and cingulate gyru) in relation to other organs of the human brain. The limbic system is located in the most secure area of the brain—at the brain core, near the brain stem, itself.[41]

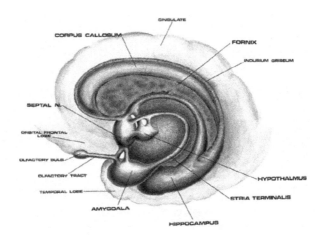

Illustration No. 2

Illustration No. 2 provides a close-up view of the separate organs of the limbic system and their relative positions and sizes. Note the complex and intricate shapes which make up the various organs of the limbic system. Overall, the limbic system appears crescent shaped, similar to that of the human ear. The illustration highlights how the limbic system is composed of an intricate arrangement of many brain organs. While the size of these organs may be small when compared to the rest of the brain, their functions and interactions are extremely complex.

The limbic system is critical to our theoretical model because it is the command and control center for all PhiAlpha processes in the dying body. We will shortly see that the limbic system's control over essential body functions, consciousness, emotions, and memories will have a direct impact upon PhiAlpha's critical functions. But before we can adequately explore the connection between the limbic system and PhiAlpha, we must first learn some basic information about each of the organs making up the limbic system.

The major organs of the limbic system are:

1. **Amygdala**

2. **Cingulate gyrus**

3. **Corpus callosum**

4. **Fornix**

5. **Hippocampus**

6. **Hypothalamus**

7. **Indusium griseum**

8. **Septal nuclei**

9. **Stria terminalis**

We will learn more about the various organs of the limbic systems as we proceed, but for now we need some basic information on each:

Both the **amygdala** and the **hippocampus** are bulbous in shape, though the hippocampus is at least twice the amygdala's size. They are both located at the base of the limbic system, immediately adjacent to the temporal lobe. The hippocampus processes new memories for long-term storage and regulates the mind's sense of present awareness.[42]

Protruding out of the amygdala is the **stria terminalis**, which carries nerve impulses to and from the septal nuclei, hypothalamic, and thalamic areas of the brain. The stria terminalis a ribbon-shaped band of nerve fibers connecting to the septal nuclei, amygdala, hypothalamus, and thalamus. It is believed to serve as a means of implanting fear and anxiety impulses upon memories.[43]

The **cingulate gyrus** sits just above the corpus callosum. Like many of the other organs of the limbic system, the cingulate gyrus is crescent-shaped and surrounds the other organs of the limbic system. It looks almost like a helmet as it sits atop the other organs of the limbic system.[44]

The **corpus callosum** is a thick crescent-shaped band of nerves connecting the left and right hemispheres of the brain. The corpus callosum is the largest nerve bundle in the brain, containing between 200 million to 250 million axons. The corpus callosum possesses the largest amount of white matter in the brain and is located between the cingulate gyrus and the fornix.[45]

The **fornix** is a crescent-shaped nerve bundle that serves as the main output channel for the hippocampus. It also carries nerve fibers to the hippocampus from the forebrain. Like the rest of the organs of the limbic system, the fornix is known to play an important role in how specific emotions are bonded upon our memories. Like the rest of the organs of the limbic system, not a lot is known about what other functions the fornix may have.[46]

The human **hypothalamus** is located at the base of the brain, near the pituitary gland. About the size of an almond, the hypothalamus is positioned almost in the exact center of the skull.

It is believed to control or mitigate a large number of brain functions, including pituitary gland function, hormonal release, circadian rhythm, body temperature, and behavior. The hypothalamus is highly reactive to light, olfactory stimuli, steroids, neurally transmitted signals from the heart and other major body organs, blood-born stimuli, infectious agents, microorganisms, and stress.[47]

The **indusium griseum** is a thin layer of gray brain matter covering the upper surface of the corpus callosum. It contains four longitudinal bundles (two on each side). Again, not much is known of its function. Its location suggests it plays an important role in the limbic system's emotional-memory organization.[48]

The **septal nuclei** are a set of structures made up of both a lateral and medial component. Bulbous in shape, they are located below the corpus callosum and are connected to the olfactory bulb, hippocampus, amygdala, hypothalamus, midbrain, cingulate gyrus, and thalamus. They are believed to play a role in pleasure responses, but the literature indicates an almost complete lack of understanding of its functioning.[49]

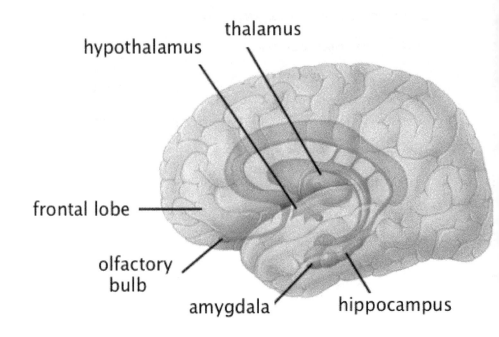

Illustration No.3

Illustration No. 3 shows the position of the thalamus in relation to other major organs of the limbic system as well as its location within the brain itself. Interestingly, this diagram depicts the limbic system as a serpent, with the amygdala and hippocampus as its head and the rest of the limbic system as its coiled body. It does not take much imagination to notice the similarity between this serpent-like shape, its connection to our emotions and our emotional behavior, and the biblical story of Adam and Eve. Is it possible this story is but an allegorical fable meant to explain to an unsophisticated agrarian culture the effect of their emotions upon their daily lives?

The **thalamus** is a knob-shaped structure of gray brain matter with twin lobes located in the center of the brain. While not considered by current medical science to be a part of the limbic system, its extensive array of nerve connections to the hippocampus suggests otherwise. It also has connections with the cerebral cortex and is believed to play a significant role in consciousness and states of sleep and wakefulness. The thalamus acts as a relay station—sorting, processing, and directing signals traveling between the spinal cord and the rest of the body.[50]

The **spinalthalamic tract** is a nerve-fiber pathway allowing signals to be transmitted between the thalamus and the spinal cord. The thalamus is believed to use the spinalthalamic tract to receive nerve impulses from the spinal cord. This connection can be better understood by examining Illustration No. 1, which shows how the spinal cord is directly connected to the midbrain and the limbic system. It is clear from Illustration No. 1 that any information going between the cerebral cortex and spinal cord must first pass through the limbic system. Later, in the chapter on memory absorption, the relationship between this limbic system and the spinal cord will provide important clues to how memories from our past lives are absorbed by our new bodies.[51]

It must be emphasized that as much as modern science has advanced in the last 50 to 100 years, it has yet to unlock the many mysteries within in the brain and the limbic system. While it is understood that the limbic system plays a critical role in how our emotions are imprinted upon our memories of specific events, the scope and specific other functions of the limbic system—and the individual organs themselves—remains largely unknown. It is a simple fact that with every single piece of knowledge we gain about the brain, we learn there is a great deal more about what the brain does that we do not yet understand.

For example, while researchers have known for some time that the brain possesses both electrical and magnetic fields, they have only just begun to study how to validate and calibrate electrical-current flow in computer models with the results taken from living human test subjects.

We have discussed how cerebralspinal fluid (CSF) is an excellent source of both electric field and magnetic field activity in the human body. The reason for this is that CSF contains many elements that, when dissolved in CSF become electrolytes such as sodium, potassium, magnesium, chloride, phosphate and bicarbonate. CSF also carries salt water and trace elements, including copper, zinc, aluminum and iron—all of which serve as excellent electrical conductors.[52]

We call the electromagnetic field generated by CSF flow the "neural net." This neural net is a biomachine created by the pulsing action of the CSF flow creating what is essentially an organic computer. As with any man-made computer, our brain runs on various programming algorithms to perform its multitude of functions.

One such function is PhiAlpha. The brain's PhiAlpha program is activated upon a significant crisis threatening our lives. Regardless of the proximate cause of our death, the initial point at which our body becomes aware of the possibility of death is what is called the "stress response," and that begins in the amygdala.[53] An almond-shaped mass of interconnected neurons located deep in the brain's medial temporal lobe, the amygdala modulates our reactions to events that we define as critical to survival.

The amygdala controls our reactions to all sorts of major events in our lives, from signaling the presence of food to warning of predators and rivals, the selection of mates to the instinctive responses to the threat of danger to children and similar crisis situations.[54] Any event the senses perceive as a credible threat to survival triggers an immediate response from the amygdala.

The amygdala's neurons may serve as the memory store necessary to confirm or reject the sensory stimuli alleging the threat. In other words, the neural memories contained within the amygdala may serve to either verify the stimuli as a valid threat or not, depending upon how the new threat stimulus matches with memories of past experiences. Certainly, to function properly as the brain's early-warning center for life-threatening events, memories contained in the amygdala would have to be very long-term memories indeed.

Upon validating a true crisis, the amygdala sends a signal to the hypothalamus. The hypothalamus is also almond shaped and is located near the amygdala in the brain's most protected core area: next to the pituitary gland, just above the brain stem.

The hypothalamus acts as the central processing unit of the body's autonomic (involuntary) nervous system. It maintains the body's internal chemical and electrical balance, and controls such functions as heart rate, blood pressure, body temperature, fluid and electrolyte balance, appetite, body weight, glandular secretions of the stomach and intestines, and sleep cycles via the release of hormones.[55]

There are few, if any, threats to the life of the body that are not detected by the hypothalamus. All these functions occur without our conscious awareness. Both amygdala and hypothalamus work together to instantly respond to crises before our brain's vision center is able to cognitively identify the event as a threat.

The following flowchart illustrates the physiological event cascade that results from the detection of any possible threat by the amygdala and hypothalamus:

1. Signal from amygdala to hypothalamus.

2. Signal from hypothalamus to involuntary nervous system (NS).

3. Signal from involuntary NS to sympathetic NS.

4. Signal from sympathetic NS to autonomic N.S. and the adrenal gland.

5. Adrenal gland creates and releases adrenaline (epinephrine).

6. Adrenaline release causes physiological changes (increased heart rate and blood pressure, and senses become significantly more focused.

7. Adrenaline is exhausted causing hypothalamus to activate secondary hypothalamus, and the pituitary and adrenal glands' (HPA axis) threat response.

8.   **HPA axis signals release of corticotropic hormone (CRH) into the pituitary gland, which then triggers release of adrenocorticotropic hormone (ACTH) that travels to the adrenal glands and initiates release of cortisol.**[56]

Cortisol is the body's last line of defense against a life-threatening event. A complex hormone released only after initial release of adrenaline has become exhausted, cortisol is the body's final biochemical stimulus to enhance performance to counter any perceived life-threatening event. Cortisol release initiates a series of physiological changes to help us survive. Some changes are easily understood. However, the purpose of several other changes created by cortisol remain mysterious.

For example, cortisol upregulates (turns on) three genes: interleukin-4; interleukin-10; and interleukin-13. These genes help the body reduce the severity of injury by increasing anti-inflammation and antibody production. As a diuretic, cortisol increases water excretion through perspiration and renal plasma flow from the kidneys.

These are just a few of the many effects of cortisol. Remember, cortisol is released only after the body has expended its entire supply of adrenaline without reducing the threat level.[57]

However, cortisol also increases sodium retention by causing the sodium-potassium ion pump to go into overdrive. When death seems imminent, cortisol forces increased retention of sodium ions at the same time it is trying to increase water excretion. Sodium atoms retain water. These processes are in opposition to each other. They conflict.

What is happening? What is so important about the cellular sodium-potassium ion pump that it forces cortisol into causing two opposing physiological processes during a time when one would logically expect all physical processes to work together, not against each other?[58]

Before we can answer this question, we must first discuss cortisol's other effects.

Cortisol has two other effects upon the body in crisis that are significant to our understanding of PhiAlpha. It stimulates the release of copper enzymes including lysyl oxidase and elastin. While copper enzymes maintain and repair connective tissue that may be damaged during the crisis, copper is also a perfect electrical conductor. By release of copper enzymes throughout the body, cortisol may be preparing the brain's neural net for the electrochemical reactions necessary to activate PhiAlpha.

This would also explain the sudden and seemingly contradictory retention of sodium ions in neural cells. Sodium ions are also an excellent electrical conductor.[59] PhiAlpha relies upon the body's electrical and electromagnetic fields to encode memories upon phased biophotons as they are released from cells within the brain.

Lastly, during extreme physical crisis, cortisol causes the brain to create what have become known as "flashbulb memories" of the crisis event.[60] We suspect these crisis memories are given this flashbulb physiological emphasis to hasten their routing to the amygdala and encoding into PhiAlpha if the crisis becomes fatal or preserved for future reference in the amygdala's long-term memory storage.

Illustration No. 4

Illustration No. 4 presents the moment of impact of a motor vehicle accident. Images such as this become part of the flashbulb album of memories the limbic system measures for the emotional weight we give them.

We conclude that it is this release of cortisol that triggers activation of PhiAlpha in the brain. Being at the heart of the body's central processing unit, the limbic system is perfectly designed to monitor the crisis and modulate the expression of PhiAlpha accordingly. In the event the body is mortally wounded, the limbic system has already set in motion those functions necessary for PhiAlpha to take over. If the crisis is averted, however, the limbic system discontinues cortisol release. PhiAlpha is shut down without the individual ever being aware it existed.

7. MEMORY ORGANIZATION

PhiAlpha is initiated once the limbic system becomes aware that death may be imminent. But PhiAlpha is a multistage process. The initial stage of PhiAlpha is memory organization.

Survivors of near-death experiences often describe a period during which they observe their lives flashing before their mind's eye. We suspect this intense flood of visualized memories—some of which are from childhood and, perhaps, even from past lives—to be direct physical evidence of this stage of PhiAlpha. We also suspect there to be a connection between the limbic system's flashbulb memories and the experience of having "life flash in front of your eyes," as has been anecdotally reported.

Biologically, PhiAlpha must perform its functions within a narrow time window. Once the heart stops, oxygen is no longer pumped into the brain through the bloodstream. Brain damage begins within three to five minutes after cardiac arrest. Brain death occurs within minutes thereafter (allowing for body temperature, causation, and extent of any physical trauma, age, chronic disease and other variables).[61]

As a metabolic process, PhiAlpha requires energy to function. This energy must come from the brain cells involved. Though even a corpse possesses a continuing ability to upregulate genes, we believe there is a bright-line, a time limit or point of no return after which PhiAlpha would no longer function due to loss of oxygen in the brain's blood supply.

Upon cardiac arrest, this energy supply is limited to whatever enzymes and other energy forms the cells can obtain internally. Once this energy supply is exhausted, the brain can no longer support the metabolic needs of downloading, prioritizing, encoding, and transmitting of memories required by PhiAlpha.

Before we discuss memory organization further, it is necessary to review how memories are made, stored, and retrieved. The issue of memory construction is relevant, because whatever the process of memory storage happens to be, we believe PhiAlpha reverses this process to organize those memories its processes identify as those necessary for individuality preservation. Science still does not understand all the basic memory processes.[62]

What is known is that there are three types of memories: ultra-short-term sensory stimuli memory; working memory for decisions and day-to-day living; and long-term memory. Events chosen for memory encoding are selected by cooperative processes of the thalamus and amygdala as they occur. At the same time the thalamus regulates neural firing of action potentials to encode the event as a memory upon a neuron, the amygdala creates an emotional imprint upon the neuron prioritizing its emotional significance.[63]

The greater the emotional significance to the memory, the higher the encoded intensity within the neuron. We suspect the amygdala controls encoding of emotional intensity by varying the strength and duration of the synaptic firing.

Memories are then stored in the appropriate sensory lobes of the cerebral cortex and regulated by the hippocampus to create a fluid memory sequence when recall of an event is required.[64] This is analogous to a motion picture. These edits are so well hidden, the observer never knows they occurred. Because emotions are unique to the individual, memories of events can vary greatly even among those who witness the same event. This is one reason testimony of even honest witnesses at trial can be very inconsistent and even contradictory. The hippocampus determines which memories are to be selected for storage as long-term memories.

How this is accomplished is not yet known.[65] The exact mechanism of memory construction is not thoroughly understood. However, it is believed to be related to electrochemical energy transfer inside the neuron itself. It is believed that a process called "long-term potentiation" allows the synapses to fire action potentials in sufficient number and intensity appropriate to the significance given the event to be placed in memory. This long-term potentiation is in the form of discharges of high-frequency electrochemical energy by the synapse or synapses involved.[66]

Additionally, one of the lesser known functions of the hippocampus appears to be the intensification of the brain's electromagnetic field at death. Experimenting with new technologies for detecting hard-to-find trace metals using laser ablation combined with mass spectrometry, Researchers J.S. Becker and Myroslav Zoriy of Forschungszentrum-Julich (Julich Research Centre), a leading German interdisciplinary research center in Julich, Germany, in 2005 found copper in higher overall concentrations in the hippocampus, including discrete copper concentrations in the dentate gyrus, and in higher concentrations in the layers of the cornu ammonis (located between the dentate gyrus and subiculum).[67]

The presence of copper within the limbic systems itself is evidence of direct involvement in the electrical conductivity needed to create an electromagnetic field.[68] But an electromagnetic field requires a continuous current of electricity to exist. Does the brain possess such a constant flow of electrons around the cortex that would link up all the memories in all the lobes into a singular, individual state of coherent energy?

The answer lies within the cerebral spinal fluid itself.

Cerebrospinal fluid (CSF) is a colorless, slightly alkaline (pH 7.28-7.32) liquid consisting of water and electrically conductive and ionizing minerals, including sodium, potassium, chloride, calcium, and magnesium.[69] The CSF occupies the subarachnoid space and the ventricular system around and inside the brain and spinal column.

This subarachnoid space contains the layers of delicate meninges tissues within which the CSF flows.

Illustration No. 5 Show how the CSF surrounds the brain like a

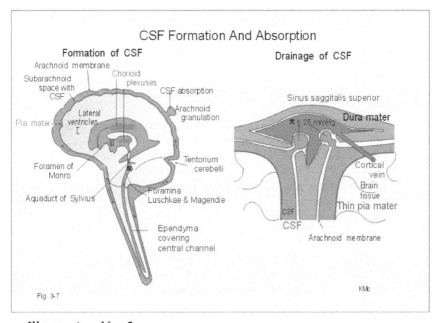

Illustration No. 5

river of electrolytic fluid. CSF is always in motion. It is this constant

115

motion of electrolytic fluid that creates the electrical and magnetic field enveloping the brain. MRI video confirms the Klorica and Oreskovic (2014) hypothesis that CSF flow is not unidirectional circulation but the result of cardiac-cycle-dependent bidirectional systolic-diastolic pulse washing forward and backward within the cranium.[70]

Darko Oreskivic and Marijan Klarica are both researchers at Ruder Boskovic Institute (RBI) at Zagreb, Croatia. RBI is the largest Croatian research center in natural sciences and technology.

Elementary particles (electrons and photons) are emitted from the transfer of ionized sodium and potassium atoms across neural cell membranes as the CSF pulsates back and forth over the cortex. This process creates a neural net of coherent information existing in a quantum mechanical state.

Evidence of this neural net was found by researcher S. B. Bauman, of the Department of Neurological Surgery at the University of Pittsburgh Medical Center. In only the second reported study of CSF conductivity, Bauman and his team found that in the seven patients sampled, CSF was found to be 23 percent higher at body temperature than at room temperature as ionic motion and diffusion increased. Further, a 3-D finite-element model of the head using more than 450,000 elements showed that scalp electric and magnetic fields are both highly sensitive to CSF conductivity.[71]

This CSF wash-cycle action across the folds of the cortex creates an electromagnetic field producing quantum mechanical effects throughout the brain—and the mind. It is proposed that this electromagnetic field creates a quantum coherence that unifies all memories together in a singularity of self-awareness—consciousness.

How is PhiAlpha initiated?

Once the heart stops, the limbic system initiates PhiAlpha as follows:

The hippocampus receives impulses from the neural net for those memories controlled by the largest and strongest synapses. These memories are released from their neurons in the form of ions released from ion channels. Ion channels are proteins that make it possible for ions within cell membranes (including neuron and synapse cells) to move across the membrane and out of the cell.[72]

Memories are transmitted instantly to the neural net created by the electromagnetic field of the CSF. The neural net then instantly transfers these memory signals to the limbic system for the next phase of PhiAlpha to begin.

It is important not to confuse this initial "download" of memory-encoded ions as the final PhiAlpha memory product emitted at death. This download is a massive release of memories that need prioritization by the limbic system's emotional control center.

8. MEMORY PRIORITIZATION

We begin this stage of PhiAlpha with several assumptions. First, PhiAlpha cannot transmit the brain's entire memory core at death. This is conjecture, but reasoned conjecture, given the limited time in which PhiAlpha must complete the processes necessary to transmit individuality by release of memory-encoded biophotons prior to brain death of the host and the amount of memory storage the brain possesses.

As discussed earlier, the brain has as many or more neural-synaptic connections as there are stars in the universe. Therefore, there must be a means by which the limbic system evaluates memories to determine which ones are to be transmitted and which ones are not.

Secondly, we theorize that PhiAlpha's transmission capability will not be the same for every individual. Physical defects of age, trauma, disease or other impairment would likely inhibit PhiAlpha from functioning as intended in each death. Further, the energy level of the biophoton burst we suspect is used to release the memories from the cranium at death would also be limited by the same variables. Coherence of the biophotons would be affected as well.

The analogy that comes to mind is a sinking ship full of passengers.

Who do we save?

Who do we leave behind?

The dying body is not a sinking ship, but the problem is the same: Which memories does the body save as PhiAlpha and which does it leave behind? How are these choices made?

For the dying individual, PhiAlpha uses the brain's limbic system as a filter to choose which memories are most necessary for preservation of the person's individuality at death. The hippocampus and the amygdala's emotion control center would most likely perform much of this task. They are both centrally located in the center of the brain, which would be among the last of the brain's components to deteriorate from oxygen loss following cardiac arrest or from whatever physical trauma the body suffered.

What memories does PhiAlpha choose?

This question forces us to examine how our individual personalities are constructed. Let us consider some examples.

When we look in a mirror and see our reflection—our emotional state is affected every time we remember our reflection. We are either happy with what our reflected image represents to or we are not. This emotional response is registered in our amygdala. This is an example of **ultra-short-term** or **sensory memory**. It will fade away in time because its impact upon the neuron carrying it is very slight compared to all the other memories contained in the cortex.

Illustration No. 6 (below) is an image of the five senses from which we receive millions of bits of sensory data every second. Sensory or ultra-short-term memory can include everything we are seeing, to the feeling of clothing and jewelry we wear, to how warm or cold we feel, and the scent of whatever odors are being carried in

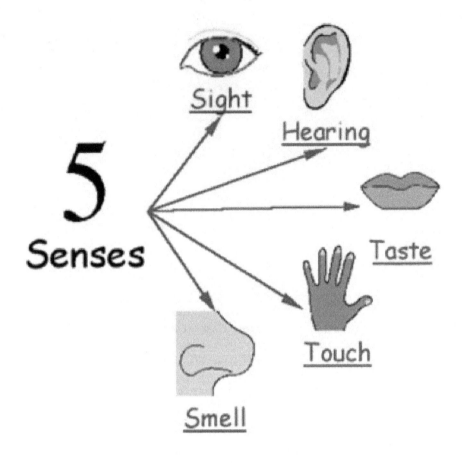

the air or cold we feel, and the scent of whatever odors are being carried in the air.

Illustration No. 6

Working memories do not last because they do not possess any emotional weight for neurons to latch onto. They are the flotsam of our day-to-day lives. They arise from the superficial contact of the conscious interaction we have with our external environment. When the drive to work is routine, there is no emotional hook upon which it can attach to a neuron. But we do remember the drive to work when we become irate over another motorist's carelessness.

The routine duties of the day can merge into one another. Each day has its own rhythm. Yet each day is so similar to every other that we take for granted most of what we do or observe. This is an example of working memory.

Illustration No. 7

Illustration No. 7 shows a typical example of what our working memory is responsible for routine day-to-day decisions such as our daily work, our daily schedules, and the subjective criteria we use to make the choices we make every day. The difference between sensory memory and working memory is based upon the question of which is more important to us: our physical environment or the activity in which we are engaged within that environment.

For example, the man in the photo above is seated at his desk where he is studying what is on his computer screen. If he is sitting on a wallet that is awkwardly placed in his back pocket, he will be uncomfortable. This is sensory memory. The report he is preparing, while enduring the increasing discomfort of that awkwardly placed wallet, is part of his working memory.

Long after he has forgotten his sore posterior, he will remember whether he was patted on the back for that report—or demoted. His brain's limbic system will register his emotional response to that evaluation on the neurons onto which the memory is encoded.

Both ultra-short-term and working memories would immediately be rejected for PhiAlpha. As the purpose of PhiAlpha is to preserve only those memories that support individuality, present sense impressions and day-to-day working memories are too immediate to carry much, if any, synaptic weight. Further, one of the means by which the hippocampus and amygdala prioritize memories is the strength and size of their neural-synaptic cells, which vary in strength and size depending upon how often they are used.

Long-term memories would logically be stored in neural-synaptic cells stronger and larger than recently created memories simply because they have existed over a longer period of time, and their synapses have been used considerably more often. This evaluation would require only a simple computational exercise by the limbic system to perform.

All that would be necessary would be for the limbic system to compare the strength and size of synapses of each long-term memory it receives. Upon cardiac arrest, ions containing long-term memories would be released into the neural net via the CSF wash. These ion-held memories would then be transferred to the amygdala for emotional prioritization by size-comparison.

A much more difficult task for the limbic system to perform would be to differentiate between long-term memories that are of equal or near-equal sizes. So, how does PhiAlpha select which long-term memories to save?

The strongest long-term memories are those memories we still recall, often without even trying. They bubble up out of subconscious like phantoms. Images of long-ago events washing over our minds like a motion picture we cannot turn off or walk away from. They seem to have a will of their own, dominating our attention regardless of what we are doing at the moment and keeping us up at night or waking us up from a sound sleep.

Evolution created PhiAlpha for the specific task of conserving information to comply with physical laws fundamental to the existence of the universe. Who each of us is as an individual is not something we can consciously identify or even observe. We are the object behind the face falling away from the canyon ledge into the mist far below—not the observer watching us fall.

Our conscious minds are held prisoner to the immediacy of our second-to-second sensory experience and the necessities of our day-to-day lives. Long-term memory creation is a program running far into our subconscious, if not farther.

If there is such a thing as a mystical component to PhiAlpha, it is at the point of long-term memory creation and prioritization where these two seemingly contradictory concepts meet. Our emotions do matter in the creation of stronger, longer-living memories.

Recent research supports this theory. It has been found that the brain's amygdala controls how the human brain preserves our emotional memories. The amygdala does this by encoding upon our memories the emotions we feel as we live through our life experiences. Emotion has both transient and long-term effects upon memory. Emotionally charged memories, including stimuli formed by pictures, words, and faces are more easily recalled over time that emotion-neutral memories.[73]

This research proves the emotional loading of our memories is controlled by the amygdala and corroborates a critical physiological step necessary for PhiAlpha's operation.

In 2003, a separate team of researchers studying the effect of negative emotions and long-term memory suggests negative emotions are the strongest and longest lasting of all memories—therefore, the easiest to recall. Further, long-term memories are given preferential treatment for recall over short-term memories.[74]

We can hypothesize that since negative memories are strongest, the synapses containing negative long-term memories have a stronger ion signature and are physically larger than other long-term memories. This preference for negative emotions in long-term memories is likely to be an evolutionary construct for survivability.

It is possible that evolution boosts the ionic signature of negative or life-threatening memories in the synapse as a means of self-preservation.

Gradual loss of brain oxygen at death limits the time for PhiAlpha to operate. All the processes necessary to download, prioritize, encode, and transmit memories must be completed before the limbic system exhausts its oxygen supply. Due to this time limitation, this process must be simpler than it might appear.

As the amygdala is the one brain organ capable of encoding emotion upon memory, it would be logical for it to possess an emotional matrix of the individual's personality—a blueprint, so to speak, of the person's identity over time (including from one life to another). This makes perfect sense given the fact that if PhiAlpha exists at all, there would have to be an organ within the brain capable of transferring emotions to memories and then encode both upon biophotons for emission out of the body.

If this is correct, all the amygdala would need to do to prioritize the ions from long-term memories is to run them over its own emotional matrix, holding on to those memories that match and disregarding those that do not. These remaining prioritized ion-held memories carrying the person's individuality-encoded memories would then be transferred to another area of the limbic system for biophoton encoding.

9. MEMORY ENCODING

Memory encoding requires three components:

1. The ions containing the memories

2. A place within the limbic system where the memories can be successfully encoded.

3. The transmission medium upon which the memories are to be emitted out of the body.

How the ion-held memories are likely prepared has already been discussed. Encoding of ion-held memories must be an electrochemical process. It may use the same principles of simple photography or something more complex such as magnetic resonance imaging (MRI). Whatever the process, it must be natural to the physiology of the brain and require little time and energy. A dying person would have little left of either. By this time, the gray matter and white matter of the brain may already be "dead" and the only part of the brain still alive may be the brain's core, where the limbic system is located.

The location within the limbic system responsible for encoding the memories for transport out of the body must be free of electrical activity that might create electrical interference with PhiAlpha encoding. Such an area would need to be structured such as to readily secure the memories in place for stable encoding. It would also need to be relatively free of cellular activity of its own—a relatively electrochemically-free zone within which the encoding could take place in the same way a darkroom is necessary to develop photographic film. Limited interference may be tolerable as cellular activity throughout the brain at this stage of the death process would most likely be greatly reduced or even absent.

One such area of the brain matching all of these requirements is the **hippocampal sulcus**; a small fissure or pocket that lies within the hippocampal formation near the amygdala. The hippocampal sulcus has the two qualities needed for PhiAlpha memory encoding.

First, its pocket-like shape can readily hold and stabilize the ion-held memories while they are being encoded upon the transport medium. Second, being relatively free of cells itself, the hippocampal sulcus provides an area relatively free of any electrochemical activity that might interfere with PhiAlpha encoding.[75]

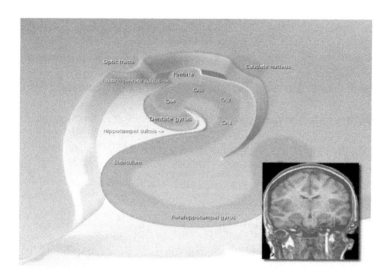

Illustration No. 8

How are memories encoded?

Illustration No. 8 shows a drawing of the hippocampal sulcus (the photographic inset is an MRI image of the brain with the hippocampal sulcus highlighted by the yellow square). As shown in the drawing, the hippocampal sulcus is shaped as an empty pocket-like fold located between the dentate gyrus and the subiculum.

The hippocampal sulcus is the brain's own little darkroom. The pocket inside the fold provides the compactness necessary to force the biophotons and the ion-held memories together. As the biophotons and memory-ions merge, there is a brief flash of light as the memories are bioelectrically transferred like a photograph from the ions to the biophotons. Once the memories are "flash-encoded" upon the biophotons, the biophotons are released and travel out of the body.

Ion-held memories prioritized in the amygdala would not have to travel far to reach the hippocampal sulcus as the hippocampus lies just below the amygdala. The ions would be held in place by the pocket-like formation of the sulcus long enough for the hippocampus to discharge a flash of biophotons from the surrounding sulcus.

This, we suspect, is the cause of the sudden, intense EEG burst observed by ER doctors treating patients who had near-death experiences. This flash of biophotons would extend throughout the surrounding hippocampal sulcus tissue, essentially creating a holographic image of the memories stored within the ions. This 360-degree flash of biophotons would have the added effect of encircling the ion-held memories in the self-contained state of energy and data coherence necessary for survival during emission out of the body.

Biophoton emission in humans has become an exciting area of study in recent years. A growing body of evidence suggests that photons play an important role in the basic functioning of cells. Cells emit light when they work and use light to communicate with each other. Photons may also play an important role in neural functioning. Spinal neurons of rats have been found to emit light. Microtubules inside the brain may act like optical fibers conducting photonic information.[76]

In addition to electrical and chemical signals propagating in the neurons of the brain, signal propagation also takes place in the form of biophotons. Experimental confirmation of photon-guiding properties of a single neuron have been documented by researchers. Interactions between biophotons and microtubules in the brain may cause transitions or fluctuations of the tubules between coherent and noncoherent states. The role played by biophotons inside the brain is an exciting new area of biophysics that merits special study.[77]

As Professor Majid Rahnama of Iran's Shahid Bahonar University suggests:

As Rhanama and his research team suggest:

"These interactions involve long-range ionic wave propagation along microtubule networks (MTNs) and AFs (action filaments) and exhibit subcellular control of ionic channel activity…"[78]

They explain how biophotons help support coherent energy states and show how the interaction of a quantized field with a system of electric dipoles produces what's called a state of coherent energy. They believe microtubules can be considered biological electrical dipoles.[79]

Other researchers have confirmed the presence of biophotons in both living and deceased human brains. Professor Joey M. Caswell and his team instructed human subjects to employ intention in effort to affect the direction of a "random number generator" within a special device located within 1 meter of the test subjects. With this device, biophoton emissions from the brain's right hemisphere were recorded simultaneously with each of the subjects' thoughts.

Caswell and his team discovered that when human test subjects attempted to affect the numbers produced by the "random number generator," their brains experienced a significant increase in photon emission. The researchers concluded that intentional thought patterns were directly responsible for these photon emissions. In other words, the researchers were able to conclude that that the human brain emits photons as a normal process of its functions.[80]

The interrelationship between the hippocampus and biophoton emission has been well documented by independent teams of scientists. Nicolas Rouleau, of the Biomolecular Science Program at Laurentian University in Sudbury, Ontario, Canada, measured biophoton counts that exceed background flux densities (electromagnetic field densities) from hippocampal tissue slices of and the brains of human test subjects engaging in specific thought processes.[81]

In a separate study, Rouleau found human hippocampal brain region specimens possessed the ability to emit biophotons 20 years after death.[82]

Lastly, a 2016 study found that glutamate-induced biophotonic activities and transmissions in the brain, present a spectral red shift from animals (in order: bullfrog, mouse, chicken, pig, and monkey) to humans, even up to the near infrared wavelength in the human brain. This increase in photonic activity in the human brain, they believe, may partly explain the high intelligence of humans because photons are a more efficient communicator than electrochemical processes.[83]

The subject of quantum mechanical processes within the brain has been an area of ongoing scientific debate for decades. But now that biophoton emission from the brain has been actually observed and studied, there can be no debate. It is proven. The brain emits biophotons continuously as part of its normal everyday function. The conclusion that the human brain is a continuous source of biophotons as the result of its electrochemical and electromagnetic processes is startling. Even more startling, however, is that this biophotonic behavior opens the door to the mind-blowing reality that the human brain is an organic quantum mechanical machine. Brain physiology creates a quantum mechanical energy state and uses this quantum energy to function in ways we have yet to even imagine.

10. MEMORY EMISSION

We are near the end of the PhiAlpha process within the dying human host. In our theoretical model, the individuality-carrying memories have now been encoded upon biophotons within the hippocampal sulcus. This packet of biophotons is now ready to be released or emitted out of the dying body. As we just discussed, biophoton emission out of the body would be controlled by quantum mechanical effects and may involve what researchers term "quantum teleportation."

In 1998, Professors A. Furusawa, J. L. Sorenson, and their team reported the first successful experimental quantum teleportation of photonic energy state.[84] In 2016, a separate research team reported successful teleportation of entangled photons over a 30-kilometer optical fiber quantum network.[85] Also in 2016, the team of Raju Valivarthi and Marcel.Li Grimau Puigibert reported the successful quantum teleportation of a telecommunication-wavelength photon interacting with another photon over 6.2 kilometers. Applications of this research in the telecommunications industry include extending quantum communication distances using quantum repeaters that rely upon light-matter (photon) entanglement.[86]

The concept of "quantum teleportation" is complicated, but its origin arises from a logical inference of quantum mechanics. The teleportation aspect of the concept is not what messes with our minds. Most of us are familiar with the concept of teleportation from *Star Trek*. It is the "quantum mechanical" component of the term that can throw us for a loop.

Quantum mechanics is a part of the universe humans lack the ability to experience with any of our five senses. We cannot see, smell, touch, hear, or communicate with the quantum mechanical part of the universe. This is so because the realm of quantum mechanics is limited to only those forces and matter so small that they are referred to as elementary particles. Elementary particles are so small they can only be studied by special machines called particle accelerators and decelerators, like the Large Hadron Collider and the Super Proton Synchrotron at CERN.

The first serious consideration of the subject of quantum-teleportation was the thought experiment Einstein, Podolsky, and

Rosen made famous in their 1935 article, "Can Quantum-Mechanical Description of Physical Reality be Complete?" Quantum entanglement is one of the weirder applications of quantum mechanics and theoretically creates, as Einstein described, "spooky action at a distance."[87]

To illustrate what Einstein meant, consider the following scenario: two particles interact and then both fly off in separate directions at the speed of light. Even when they are so far apart that classical mechanical laws no longer apply, a quantum mechanical measurement of either particle will instantly provide descriptive information about the other.[88]

This action is described as spooky because it implies that with entangled particles, information can travel instantly, regardless of the distance separating them—in apparent violation of the laws of classical mechanics stating nothing can travel faster than light itself (186,000 miles per second). There is nothing spookier than the idea that quantum entanglement allows light waves to travel across intergalactic distances instantly.[89]

Several questions pose themselves at this point regarding PhiAlpha emission. How much information can one photon carry? How many photons are needed to encode PhiAlpha memories? How many biophotons could be stored in the hippocampal sulcus and emitted together as discrete PhiAlpha before emission? How would this PhiAlpha matrix be emitted; as a collection of joined particles, or as a wave? Where would PhiAlpha be transmitted, and how would it get there?

These are all critical questions worthy of scientific interest. However, without an afterlife science, these important issues will remain unanswered indefinitely.

We should review some basic information. Photons are elementary particles that have zero-rest mass and are always traveling at the speed of light. Like electrons and other elementary particles, describing photon behavior (including biophotons) is restricted to the abstract logic of quantum mechanics. Photons are subject to what is termed "wave-particle duality." Photons share the properties of both particles as well as waves.[90]

However, the famous physicist and Nobel Laureate Richard Feynman believed both the wave viewpoint and the particle viewpoint were inaccurate. An accurate description of photonic particle and wave properties is very complex. While this subject extends far beyond the scope of this book, his lecture on the topic, archived at the California Institute of Technology, is well worth studying.[91]

We know from our earlier discussion of quantum entanglement that photons can be measured as entangled particles. We also know photons are measured in the electromagnetic spectrum as waves whose wavelengths vary with frequency or energy of the photon (which determines the colors we see in the visual spectrum). So, let us just accept as true that photons are both particles and waves relative to how we interact with them.

Measuring individual photons is extremely difficult. Like any elementary particle, photons are fragile and weak. Merely measuring them can be enough to interfere with their properties.

Photons of PhiAlpha would be emitted from the hippocampal sulcus either as discrete particles or as an electromagnetic wave or series of waves. They would most likely then absorb the ions containing PhiAlpha memories and immediately exit the dying or dead body as a wave of radiation with a specific frequency and wavelength. If correct, this emission of photonic radiation would then be measurable even with current technology. It would most likely be one sustained pulse that travels as a wave, though its frequency and wavelength are presently unknown.

A promising starting point for further research in this area should focus upon the correlation between the gyrus/sulcus fold-wave pattern and photonic wavelength. We do not believe it possible for neural net ions to travel from neuron to synapse to neuron through wavelike axonal channels without absorbing an energy signature matching the wave-path they traveled through the cortex by.

However, it is possible that the simple act of measuring PhiAlpha might be enough to destroy its energy coherence—in effect, destroying the memories it contains and the individual personality they are attempting to preserve. Even measuring elementary particles may be enough of an interference to affect their behavior. Therefore, it is important to create instrumentation that will be able to observe

PhiAlpha without adversely affecting the individuality matrix contained within.

11. MEMORY RECEPTION

So far, we have shown the steps needed for our individuality to be transferred out of our dying body: our individuality is downloaded in memory-signals from our neurons to the limbic system for prioritization. The memories are then transferred by ions to the hippocampal sulcus for flash encoding upon biophotons, forming a biophotonic packet containing the memories. The biophoton packet is then emitted out of the brain.

Where do biophotons go?

Light acts as both a particle and a wave. As a wave, PhiAlpha can be both observed and measured. PhiAlpha will have a wavelength and frequency just like any other electromagnetic wave. PhiAlpha will be capable of being received—just like any other wave—given the appropriate receiving antenna.

There are only two logical options as to what this receiving antenna could be. Either the receptor is a biological life-form or it is not. Simple logic dictates the conclusion that evolution intended for PhiAlpha to be received by the same species from which it was transmitted. Evolution is, by process and purpose, a biologically oriented process. Therefore, it can only transmit information from one life-form to another.

Evolution crafted PhiAlpha as an electromagnetic wave to be received and absorbed by a fetus in the womb as quickly as possible after emission and release from the dying host body. Any delay could result in PhiAlpha's energy coherence decaying, threatening the viability of the information stored inside. As PhiAlpha is most likely an electromagnetic wave, it is possible it may be intercepted, relayed to an intermediate storage area, and tampered with. This unnatural intrusion into the evolutionarily created and moderated procedure would threaten the very underpinnings of biological life itself.

We will restrict our discussion here to the most likely method of transmission: that the PhiAlpha matrix is transmitted as a quantum mechanical state and received by a human fetus through the antenna-like organ of the spinal cord before birth. This is not as strange as it may sound.

California State University, Sacramento, Professor S. Balaguru's paper on "Investigation of the spinal cord as a natural receptor antenna for incident EM waves and possible impact on the central nervous system" reported his team's research on the effects of EMF waves on biological systems including the spinal cord.[92] Their application of modern modeling technology known as finite difference time domain (FDTD) was used to simulate various voltages and electric currents in different parts of human and animal tissues. When they applied this technology to analyze incoming EMF waves to the spinal cord, they observed the spinal cord acts as a natural antenna.[93]

The spinal cord begins to develop with the rest of the brain at week five (three weeks post-conception).[94] Therefore, it appears that both the brain and spinal cord grow together from the beginning as one system. In fact, images of a 3-month-old fetus shows the spinal cord already developed as a fibrous filament that enters the brain and merges with the pons and cerebral peduncle as one undifferentiated tissue. At this stage, the spinal cord is shaped like a small thread or filament. The thread-shaped tissue of the spinal cord extends into the brain, curling like a snake around the thalamus into the midbrain itself.[95]

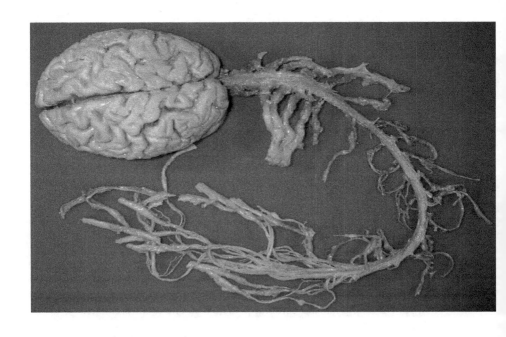

Illustration No. 9

Illustration No. 9 is a photograph of a complete human spinal cord and its connection to the brain. Special note should be given to the length of the cord and its antenna-like appearance.

Though the main part of the spinal cord is located within the spinal column down the back,

its nerve-roots extend like tendrils outward from the spinal cord. Together they create a web-like structure perfectly suited as a bio-antenna to receive the PhiAlpha wave signal.

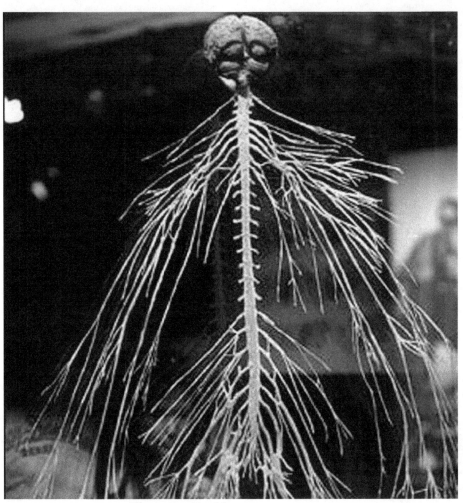
Illustration No. 10

Illustration No. 10 shows how the human spinal cord's nerve roots look when spread out, as they would be in the body. This illustration shows how similar the extended nerve roots are to many modern web-mesh antennae arrays used in telecommunications.

Like any other electromagnetic wave, PhiAlpha passes through the mother's tissue and is absorbed by the fetus's developing spinal cord like any antenna absorbs electromagnetic radiation of receivable wavelength. The size, length, width, and density of the cord and nerve-root tissue may all play a role in determining the wavelength and frequency required for reception of the PhiAlpha wave signal.

Once implanted within the tissue of the spinal cord, PhiAlpha memories are gradually unencoded from their carrier photons. The spinal cord tissue would separate the photonic particles from the ionic particles. The photons would be absorbed as energy, leaving the ions behind for decoding by the fetal limbic system when they were developed enough to perform that task. Until, then, however, the ionic memories would remain, stored safely within the spinal cord.

How long PhiAlpha can exist as a phased state of coherent, disembodied energy is unknown. In other words, we cannot say with certainty how long the emitted biophoton packet of individuality-encoded memories can remain viable without absorption into fetal spinal cord tissue. Without some device with which to artificially maintain congruity of the phased state coherence, we do not believe PhiAlpha could maintain the coherence of the photonically embedded memories indefinitely.

At first glance, the concept of our spinal cord tissue being an electromagnetic communication device sounds incredible. Nevertheless, the idea of a fetal spinal cord acting as a virtual radio wave antenna is not only possible, it is highly likely. Examine the following three illustrations as examples of biological and human-made antennae:

Illustration No. 11

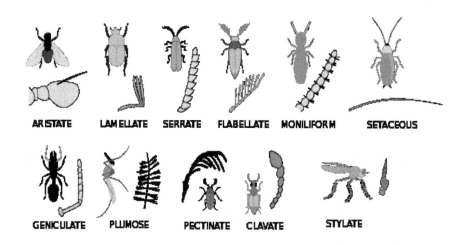

Illustration No. 12

Illustration No. 13

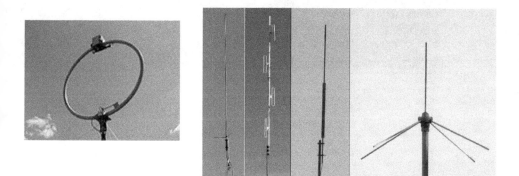

As the above illustrations demonstrate, biological and artificial antennae come in many shapes and sizes. In fact, many man-made antennae are modeled or copied from those of nature.

Illustration No. 14

Illustration No. 14 is a satellite voice antenna used by the United States military. Its wide frame mimics the web-like nerve roots depicted in Illustration No. 10. They both have a connecting center rod (in Illustration No. 10, this rod is the spinal cord) with horizontal extensions (in Illustration No. 10, these extensions are the nerve roots). While each looks different, the principle used is the same in each—reception of an electromagnetic signal.

12. VALIDATION TESTING

Validation testing is the technical term scientists use to describe the experimentation done to determine the legitimacy of any scientific theory. Why a chapter dedicated solely to this issue?

PhiAlpha is a mind-stretching theory. This is the first articulated description of a nonreligious alternative to the afterlife in human history. Sit back and ponder that a moment. How often do you have the opportunity to read and study something no one else has done—ever?

Now that you have had a moment of inspired contemplation, forget about it. Because there's a problem. The problem is proof. How do we go about proving PhiAlpha is real? All this book can do is point to the research that has already been done on unrelated topics and how that research paints a picture called PhiAlpha. All this book can do is point the way to what other research is possible to further explore whether the theory of PhiAlpha holds or not.

What follows is not meant to be an exhaustive list of the means by which we may now test PhiAlpha's validity. In fact, there is very little we know about many of the brain's organs—especially those within the limbic system. It may be years, even decades, before we are able to see the brain in its wholistic entirety. Until then, we continue to struggle for a picture of the forest while lost among the trees.

But to every scholar and scientist and researcher who doubts the existence of PhiAlpha, it is hereby demanded: before you argue against PhiAlpha's existence, first find evidence that it does not exist. For, in this book, is compelling evidence that it does.

There are at least three ways of researching the existence of PhiAlpha that do not involve active capture of a PhiAlpha packet: (1) passive monitoring of the dying for biophoton and/or ion emission; (2) passive monitoring of a fetus for biophoton and/or ion reception; and (3) genetic research.

Clearly, active capture of a PhiAlpha packet (which is not possible with present technology) of biophoton-encased ions carrying a human memory download at death would involve significant legal, ethical, and moral issues making such an effort untenable. Therefore, we must be extremely careful how we proceed. Passive-monitoring (which is possible with present technology) is clearly the safest approach and ensures no interference with the PhiAlpha packet during testing.

As we plan our experimentation program, we must consider what the possible impact upon the PhiAlpha packet would be if the theory is valid. How would the individuality preserved by the ion memories encased in the biophoton packet be affected by our testing techniques? As we monitor the subjects, we must always remember to place ourselves in their position and assume the responsibility as guardians of their preserved memories, which may be all that remains of their consciousness and individuality. It is, indeed, a most sacred trust.

In response to those who contend that with death the deceased no longer possesses any legal standing or right to complain about what happens to what is now a corpse, let us remember that our purpose here is to test the validity of a theory of possible immortality—indirect though it may be.

If one's memories are immortal, then is not the one from whom the memories originate immortal as well? Can we say it is impossible for one's future physical self to remember what was done to one's memories during the transition from dying host to fetal host? These are both reasons why passive monitoring for possible biophoton emission at death is the only acceptable validation regime.

It is not necessary to capture a PhiAlpha packet to validate the theoretical model. Technology exists permitting the passive monitoring of a dying subject's biophoton emission. Such a program would involve a human test subject dying in a controlled environment (hospital ICU) while researchers monitor electronic equipment, such as single-photon detectors attempting to observe one or more photons, photon packets or ions emitted from the dying subject. Detection of a photon or ion emission, under controlled conditions, would be compelling evidence that PhiAlpha exists.

The past few years have seen a rapid development of single-photon detectors (SPD) for passive monitoring including photon-number-resolving single-photon detectors such as: (1) transition-edge sensors (TES); (2) superconducting tunnel junction detectors (STJ); (3)

parallel-superconducting nanowire single-photon detectors; (4) charge-integration single-photon detectors; (5) visible light single-photon detectors; (6) quantum-dot optically-gated field effect transistor detectors; (7) time-multiplexed single-photon avalanche detectors; and (8) single-photon detector arrays.[96]

Photon-number-resolving SPDs are capable of precisely counting the number of photons emitted that makes this type of single-photon detector our preferred choice over detectors that do not count photons. We want to know the precise number of photons needed to encase the ions we hypothesize are encoded with memories at death. As you may recall, the photon-packet is necessary to ensure continued coherence of the ion-memory core as it leaves the dying body and transitions freely as an EMF wave to a receptive fetus.

This leads us to the many technical issues that will have to be resolved during testing. We do not know (and will not until we begin testing) how many biophotons are needed to transport PhiAlpha memories from a dying host to the fetal host. Nor do we know at what wavelength the biophoton(s) will be emitted. Will the emitted biophoton be in the ultraviolet? Will it be in the infrared? We do not know.

As a result, we will need a detector that can be calibrated for as broad a wavelength coverage as possible. If that is not practical, we will need to employ more than a one detector to ensure all possible electromagnetic wavelengths are monitored during testing. Certain wavelengths will create their own problems. For example, near-infrared and mid-infrared wavelengths require narrower nanowires (20-30 nanometers wide)[97] and lower superconducting critical temperatures.[98]

As challenging as the scientific and technical issues will be to resolve, they will not be as crucial to the project as will be those issues relating to the use of human subjects. We are convinced we now have the technology to test our theoretical model's validity and that all scientific and technical issues that arise during the program will ultimately be resolved.

The other two means of testing existence of PhiAlpha are:

(1)　attempting to measure a biophoton packet of memory ions being absorbed by a fetal spinal cord.

(2)　attempting to locate genetic markers of PhiAlpha through genetic research.

Each of these have their own advantages and disadvantages as practical research methodologies.

First, regarding fetal-spinal-cord absorption, we lack knowledge as to exactly when this occurs. Does it occur in the womb or after birth? Our preliminary hypothesis is that absorption probably occurs in the womb and that downloading memories from the spinal cord into the fetal neural net occurs gradually and may well continue into early childhood.

We hypothesize that fetal absorption of memories would serve as the "on-switch," jump-starting the process of increasingly complex associations within the fetal brain (ultimately resulting in a critical mass event of self-awareness we call consciousness). What we would look for is some evidence that the fetus is exhibiting planned behavior evidencing this growing perception of self.

Evidence of this growing awareness may be found in what some mothers report as fetal movement within the womb (called the quickening or fetal flutters). These movements often begin as early as 12 to 16 weeks from the start of the mother's last period. During early pregnancy, these movements are irregular and show no sign of intention.

However, at some point during the third trimester, these movements take on a clear regularity and can be timed by the mothers. This pattern is apparently unique to each unborn child, so mothers are often asked to time these movements to become aware of their baby's individual rhythm. Clearly, the regularity of the movements is indicative of intention and, therefore, evidence of self-awareness.

The problem is that to we do not know how much of a download of memories from spinal cord to neural net is required to initiate self-awareness. It could be immediate—or it could take weeks or even months. At this point, we just have no way of predicting the timeframe involved. Until this problem can be resolved, monitoring the expectant mother during the entire pregnancy will pose its own issues.

Our third option is genetic research. However, as we discussed earlier, PhiAlpha necessarily involves interaction between multiple biological systems in the human body, likely coordinated through the limbic system. These interactions may not be easily deduced or discovered genetically due to the indirect nature of PhiAlpha processes.

Further complicating detection, we believe, is that PhiAlpha may itself be a near-death upregulation of genetic markers that lie dormant during the rest of one's lifetime. If this is the case, we fear detecting genetic locales where PhiAlpha processes have awakened will be extremely difficult to detect until the precise moment of death.

As a result, we conclude that the testing program that provides the greatest opportunity for a scientifically valid result is the program of monitoring a subject who is near death. If pursued consistent with the concerns addressed herein, we believe such a program offers the best chance of humanely obtaining evidence of whether PhiAlpha exists.

The most critical factors of this program will be selection of suitable subject candidates and a test location. These two factors are interrelated. Due to the research involved, all test subjects must be individuals who are expected to die in the near future. Their care and comfort must be the most important consideration. Informed consent will be required for every candidate in addition to at least one, if not two, family members. No candidates should be accepted into the testing program (no matter how motivated they may be) unless they enjoy the full and active support of their families.

Candidate evaluation and selection is, therefore, of critical importance. The testing program requires that nonfamily professionals be present at a person's death—the very time families want privacy and seclusion. Therefore, consideration must be given, not only to the physical, emotional, and spiritual well-being of the candidate but also that of his or her family. Resistance or uncertainty by either candidate or family concerning the test process should disqualify that candidate from participation in the program.

Test candidacy is a decision, not just for the potential subject but for his or her family as well. Written consent must be obtained from family as well as the candidate. Candidates and their families will also need to be notified of all known possible risks, including that their biophotons may be absorbed by the electronic monitoring devices during monitoring. The risk of electron absorption during testing has been noted in recent literature.[99]

To ensure candidates are mentally competent to agree to be a test subject, each should be given a thorough psychological/psychiatric evaluation at the beginning of the evaluation program with regular follow-up evaluations as time of death nears to ensure the subject retains competency to withdraw his or her consent until the end. Any evidence of mental incompetency should immediately disqualify the individual as a test subject. In addition, no one should be accepted as a test subject who has an appointed legal guardian, guardian for medical decisions, or has allowed anyone to have power of attorney authority. Any of these should automatically disqualify a candidate.

Mental competency and informed consent laws vary from state to state. Care will need to be taken to ensure compliance with all applicable statutes and governmental rules promulgated to regulate these issues. This is another reason why universities are the preferred structure for this program. Most major research universities have legal departments experienced in handling the complex human-test-subject legal and ethical compliance issues. These universities also have research hospitals capable of facilitating all necessary program requirements. Finally, it makes logistical and organizational sense to employ a university whose physics department and research hospital can coordinate the actual testing.

As candidate and family interviews, evaluation, and preparation will take substantial time, it will be necessary to select as potential test-subjects individuals who fit the following pre-screening medical criteria: (1) chronically ill for over a year; (2) terminal diagnosis at least six months prior to screening; (3) no longer expecting a cure and resigned to death in the near future; and (4) a desire to share one's last hours for scientific research for the betterment of humanity.

Can we expect an abundance of willing candidates? There will be those for whom such a sacrifice of privacy at so delicate a time will be too much to accept. However, due to the historic scientific and social significance of the research, suitable subjects may not be difficult to find.

There will be many more who see being a test subject as an opportunity to do something great and noble in their last hours. They would have the comforting knowledge that their act of selflessness helped humanity grow in self-awareness.

13. CONCLUSIONS

PhiAlpha is a theoretical model of how evolution may have responded to forces of nature compelling it to conserve information. We have described PhiAlpha's stages, from evolutionary origin through each stage of its process, and have explained how these separate stages come together to operate as PhiAlpha. Lastly, we have provided the reader cited references establishing each stage of PhiAlpha as scientifically valid. Every stage of the PhiAlpha process has been described, explained and its application shown to be both biologically valid and corroborated by current science.

PhiAlpha is the first articulated theory of how evolution most likely adapted human biology to attain a form of immortality through the preservation of individuality-encoded memories. PhiAlpha is presented to encourage academic and public debate upon the possibility of immortality through evolution. This book is also dedicated to the need for a serious, scientific analysis of the potential for life after death.

This book does not prove PhiAlpha exists. But, from what we have observed through the analysis of our theoretical model, neither can we say PhiAlpha does not exist. In fact, this book presents sufficient evidence of biochemical and quantum mechanical feasibility in PhiAlpha processes that we must conclude there is a probability that PhiAlpha does exist.

The most compelling argument on behalf of PhiAlpha's existence is that it just makes pragmatic sense. Evolution is built upon survivability and heritability. If a need arose to require survivability of a genetic trait such as memory, evolution would create a means of doing so and protect it from undue tampering or interference—just as it created gyri and sulci to increase cognitive awareness.

When examining the merits of a criminal case, trial lawyers analyze the accumulated evidence for what we trial lawyers term "reasonable suspicion" and "probable cause." These terms are expressions we, in the legal community, use to measure probability a crime was committed and the suspect was the one who committed the offense.

As the term implies, "reasonable suspicion" suggests a conclusion that sufficient evidence exists to require further investigation; such as a voluntary encounter, stop and talk, or even a consensual search or voluntary interrogation. These are preliminary methods law enforcement utilizes to gather more evidence against the suspect. These methods may or may not yield further evidence.

However, if more evidence of guilt is obtained, the level of certainty that the suspect may have committed the offense in question may rise to the level of "probable cause." A conclusion of "probable cause" is a legal conclusion made by a judge, prosecutor, or law enforcement that the probability a criminal act perpetuated by the suspect did occur is greater than 50 percent, or, to put it another way, the odds a crime occurred are "more likely than not." In a court of law, probable cause is sufficient to justify arrest and prosecution.

Where should we place PhiAlpha on the gradient of scientific certainty at this point? At this point, PhiAlpha is but an untested theory. Yet, as this paper amply provides, there exists enough evidence of PhiAlpha to conclude there is reasonable suspicion that some form of biological process is at work that operates at death to preserve information, enough that testing should begin in earnest as soon as possible.

With PhiAlpha, death is not an end of life—and never was. With PhiAlpha, energy and information are conserved through transformation via biochemical, thermodynamic, and quantum mechanical processes allowing our individuality to survive our "physical" death.

We cannot say whether our human host bodies were ever meant to be immortal. Based upon our discussion here, however, we can say our individuality (personality) was meant to be as immortal as the laws of conservation of energy and entropy (information) allow. The only limitation would be the sufficiency of human host bodies (now being reproduced at will). As long as humanity survives, there is continuity of individuality—through PhiAlpha's preservation of memory.

Research into PhiAlpha's validity can be accomplished without risking capture or other interference with PhiAlpha particles. We recommend no one attempt to measure or capture PhiAlpha particles until these issues are thoroughly investigated and there is consensus that such efforts are harmless to the identity being transported. PhiAlpha is life at its purest but also at its most vulnerable and evanescent. It is a sacred life-form. No one has the right to interfere with its existence.

There will be those who fear PhiAlpha threatens their religious beliefs. Evolution has been developing PhiAlpha long before we humans were created. We are its progeny whether we like it or not. What force or forces set evolution in motion has us firmly in the palm of its hand. Our approval does not appear to be necessary or even requested.

However, all religions share common values and concepts that are also inherent within PhiAlpha: such as (1) karma; (2) thoughts are things; (3) what we do to others will come back to us tenfold (in memories and their impact upon our individuality); and (4) the kingdom of heaven lies within us. Each of these is consistent with the best of every major religion and PhiAlpha. Perhaps humanity's embrace of spiritual meaning is but a subconscious echo of the PhiAlpha within each of us.

PhiAlpha is an instrumental component of life. PhiAlpha is the transfer of individuality from one life-form to another. PhiAlpha has a sacred, spiritual context independent of our religious ideologies. It is quite possible our ancient ancestors were instinctively aware of PhiAlpha but gave it a religious label out of ignorance and/or misplaced allegiance to their priesthood.

Life is a gift beholden to no one and, therefore, sacred to all. Religion sees science as a threat to its power base and rightly so. Spirituality, however, has no reason to fear science. In fact, the two are mutually reinforcing. As Carl Sagan expressed so eloquently:

"Science is not only compatible with spirituality; it is a profound source of spirituality. When we recognize our place in an immensity of light-years and in the passage of ages, when we grasp the intricacy, beauty, and subtlety of life, then that soaring feeling, that sense of elation and humility combined, is surely spiritual. ...The notion that science and spirituality are somehow mutually exclusive does a disservice to both."[100]

Carl Sagan (1997)

Since our earliest humanoid ancestors stared up into the heavens looking for their celestial reflection, the universe has bestowed upon them and their progeny an infinite variety of wondrous gifts and mysteries, each one yielding its singular piece of the puzzle that is the fabric of the cloth we call existence. PhiAlpha presents one more piece of fabric for us to wear.

PART TWO:

PHIALPHA—SPIRAL STAIRCASE OF THE SOUL

14. PHIALPHA

Given what has come before in Part One, we must now assume that PhiAlpha is valid. Let us also assume that PhiAlpha's emergence is the result of evolution's adaptation of human biology to conserve what is uniquely human—the ever-expanding trait of individuality.

The following conclusions then become irrefutable:

1. Since the dawn of humanity, PhiAlpha has been present within our biological processes, preserving the memories that define us as individuals and then passing them on to each successive generation. We are the children of those memories. Our lives reflect the personalities—both good and bad—developed over those many lifetimes.

2. We carry with us from lifetime to lifetime a memory matrix of all our past lives. This can be nothing else but what is referred to as the soul. This soul is not mythical but a physical structure, however small, which must obey natural laws. The superficial projection of the soul upon physical reality is the self we are consciously aware of

during day-to-day life. In other words, the self is the reflection we see when we look into the impenetrable depths of our souls.

These assumptions are not inconsistent. PhiAlpha lends physical validity to the existence of a soul and its existence as an eternal storehouse of individuality consistent with the law of conservation of information within the second law of thermodynamics. We can see the soul for what it is, not as a mystery religion cannot explain but as a physical reality—as memory-encoded biophotons that transfer human individuality from one life to the next. Thus, the human soul and PhiAlpha are intrinsically intertwined. They are the same.

Spirit is the carrier-wave of the soul. Spirit is the coherent energy matrix that binds each soul together with one another and as part of the universe. The soul is the eternal essence of human individuality which we create from life to life. PhiAlpha is simply the biological process that evolution has crafted to ensure continuity of the soul between each life.

Spirit exists as vibrating waves of universally harmonious energies that fill the soul with knowledge, love, compassion, inspiration, and creation. The soul is a transcendent instrument of spiritual enlightenment which reacts to this spirit. Each of us has been given a soul to play like a piano as we compose symphonies of living music with each of our lives.

What makes one soul different from another? The answer is that each soul becomes increasingly unique from each of the others by the memories each accumulates during the multiple lifetimes each one experiences. Each lifetime adds a new layer of memories upon those that came before. Each lifetime further differentiates each soul from the others.

PhiAlpha does not exist to destroy religion. PhiAlpha was not created to attack religion. Evolution and PhiAlpha existed long before and independently from religion. It is not PhiAlpha's fault that it conflicts with Christian teaching. It is the fault of Christian "leaders" who chose to build their "temples" upon ideological sand.

PhiAlpha cannot be prayed away or excommunicated into nonexistence. PhiAlpha has always existed—it exists now, and it shall exist into infinity—despite any resistance from Christianity or any other religion.

While PhiAlpha benefits from an internally consistent logical flow, it shall remain only as a starting point for actual scientific research. It is only a theory. Future discoveries may reveal PhiAlpha to be more, or less, correct than other theories that will eventually be developed. What is important, however, is not to let this beginning become lost in the noise of a substantial religious backlash.

Some will argue that PhiAlpha is a threat to religious doctrine in general and Christianity in particular. They will claim that PhiAlpha challenges long-held biblical beliefs, among them the existence of a literal heaven and hell; that Christians should not believe in any form of reincarnation; whether one goes to heaven or hell depending upon baptism, confession of sin, and faith in Jesus Christ as savior.

It will be very difficult for fundamentalist Christians to adapt to a world in which life extends beyond the grave through PhiAlpha—an evolution-originated process. Their religious indoctrination has turned their minds into an intellectual prison—nothing new can enter, and they cannot escape. Until something radical and profound enough comes along to shatter those concrete bonds, they are lost.

PhiAlpha may be just what they need. One scientist from a well-known religiously affiliated American university made the startlingly unscientific claim that PhiAlpha is not science because it can never be proven. This scientist's emotional reaction reveals the extent to which PhiAlpha threatened his religious-ideological foundation.

This is what happens when an otherwise intelligent, highly educated person allows his religious indoctrination to overwhelm his scientific discipline. No true scientist worthy of the profession would ever use the word "never" concerning a theory. Technology's continuing advancement has proved the point over and over again. Today's unproven theory is often tomorrow's reality.

If this "academic" thought about what he was saying, he would be ashamed of himself. Until recently, gravity waves were purely theoretical. We lacked the technology to prove they exited. But, I do not recall any academic claiming the study of gravity was a waste of time. Until the mid-1900s, we could not prove $\mathbf{E = mc^2}$, but that did not stop American and German physicists from trying to build atomic bombs based on only Einstein's theoretical computations.

So, we must live with PhiAlpha, whether we like what it implies or not. We cannot not make evolution go away simply because its implications are inconvenient or because it challenges our preconceived notions of religious doctrine. PhiAlpha, like any other theoretical model is based upon what logical inference suggests may be possible—not what is obvious or necessarily apparent.

Despite Carl Sagan's quote in the previous chapter arguing that science and religion can coexist, there are those for whom every new scientific revelation is but a threat upon their cherished religious beliefs.

How will the concept of human immortality through evolution (PhiAlpha) be perceived by Christians? For that matter, how will PhiAlpha be perceived by Jews, Muslims, or Hindus? This will, of course, depend upon a combination of one's own religious indoctrination, how one's life experiences have shaped those initial teachings, and the religious community with which one is affiliated.

We are still only in the beginning stage of humanity's spiritual enlightenment and, as a result, we can expect a variety of reactions across the entire spectrum of understanding. Of course, the more one tends to believe in a strict orthodoxy-based religious belief system, the more likely one will object. This is only logical, as it is fundamental religious belief that will be at greatest risk from a unreligiously controlled path to the afterlife.

As understanding of the evolutionary and biological of PhiAlpha grows, the more fundamental and conservative Christians may see PhiAlpha as a threat to the relevance of their religious beliefs. As Americans increasingly seek to merge their religious views with political parties and candidates for elected office, their resistance will likely become politicized—just as evolution has become.

This resistance, a result of fear and ignorance, would be both unfortunate and unnecessary. True Christians and those who are sincere in their spiritual humility regardless of their religious upbringing and indoctrination, will realize biologically based immortality does not mean the death of their God. They will understand the underlying premise upon which religion must be judged:

God must be separate and apart from its creation—and act independently. The gardener is not the garden, and the garden does not define who the gardener is. The garden is merely a reflection of the gardener, nothing more. Therefore, humanity is not limited to the essence of God alone, but is free to grow beyond its design—and its designer.

But only if individuals choose to do so. Are we willing to accept that responsibility?

The concepts of godliness that we have developed within our own minds and shared with each other are not dependent upon the man-made rules of religious organizations whose primary function is to make money and wield political and financial power and influence for the sake of their leaders' material wealth and egos.

For all we know, genetics is a gift from God. If true, then God also created PhiAlpha. With PhiAlpha, creationism can no longer exist without science any more than it can live without God. If God created the universe, it also created the physical and biological laws that rule its processes. They are inseparable. Any other conclusion is illogical and irrational. PhiAlpha permits no other conclusion.

Remember what PhiAlpha does. PhiAlpha preserves the unique information that forms the basis of our personalities and individuality by creating a safe harbor for the memory-encoded biophotons of the dying until absorbed in the tissue of a fetus.

Wouldn't God want us to be individuals? Didn't God grant humans free will? If God did not want us to be individual aspects of creation, then God would not have created individualized souls. Further, if God respects the concept of free will then it has no choice but to expect us to develop our own individuality and respect the consequences that free will permits—both good and bad. There is nothing ungodly in that.

If you are a gardener and you plant an apple tree, you must expect it to grow apples. If you plant the apple tree, you accept that you will get both good apples and bad. You leave the tree alone. You don't kill the tree off just because it produces some bad fruit or because some of its branches grow in unpopular shapes and sizes. If you don't want to deal with bad apples, don't plant an apple tree.

Does anyone contest the concept that our souls were each meant to be unique from one another? Is there anything more sacred to

Christians than the precept that each soul is unique to its host? Is not one's soul truly one's own? Can I possess your soul? Can you possess mine?

The answer is obviously no. Each soul can only be owned by an individual human. So, the concept of the soul, itself, must be based upon the creation of spiritual individuality manifested by the human spirit and body. This tenet of soul individuality is one that Christians worship as a strongly held belief of their religious doctrine. To believe otherwise would be to allow one person's actions in life to determine the spiritual fate of another.

Individuality of the human soul is a gift, not a curse. True, the attainment and development of one's own individuality is one of the most difficult processes we humans face during our lives—but also the most rewarding.

Most see their own individuality as a priceless artifact to be treasured, nurtured, and protected at all costs. For every person I have met who expressed frustration over their interaction with others and/or the world, I have met a thousand who loved who they were as individuals. Though, for some of them, it was the rest of the world that was at fault.

So, any god creating a process that preordained the origination of individuality would logically expect the beings borne with this capacity to be individuals who would cherish the gift. The soul, whose very nature is individualistic, would not wish to live its life in the pursuit of its blossoming individuality only to lose what it had gained, at such a high cost, just because its host, the physical body, died.

Humanity has evolved over billions of years from one-celled organisms to what we are today. That is a long time to wait for a tree to grow—even for a god. If individuality is the main fruit of this Tree of Life, would God allow its value to be wasted simply due to the transiency of human life? That would be internally inconsistent, and internal inconsistency cannot logically exist within the same reference frame of any given system. No system can pick and choose which laws to obey and when to obey them. Continuity of expression is a fundamental precept of our universe's natural laws.

The concept of PhiAlpha goes beyond the conservation of unique information required by the second law of thermodynamics. The science of PhiAlpha is founded upon an inherent logical consistency. There is no logical purpose to the origin and development of individuality without provision for preserving its end product—individuality's ever-increasing complexity of expression.

Religious fundamentalists believe creation took six days. If creation was completed within the first six days of existence, what has been happening ever since? To believe creation ended at the Big Bang is to believe the first chapter of a suspense novel is the climax, or that a Broadway play ends after the first scene or that a football game is over right after the kickoff.

We humans tend to jump to conclusions without having enough data. We think we know all the answers, but often we don't. So, we make mistakes that could have been avoided.

Creation of individuality without the provision for continuity would be like creating the laws of physics but denying the universe the physical ability to form planets, solar systems, and galaxies consistent with those laws. The consequences would be internally inconsistent with the causal processes. This would be neither logical nor rational.

Religious faith is not a logical process, however. People who act upon their religious faith do not do so as a result of what they observe or perceive as objective reality but often in spite of it. Their actions are the result of their inner voice and the subjective emotions that arise out of it. Conscience exists independently of external forces. The force within has greater influence upon behavior than any external stimulus. The more one's life is dependent upon one's religious faith, the more likely he or she will feel threatened by PhiAlpha.

The reality is that our religious growth is being stunted by the institutions that depend upon our religious obedience for money and power—the church. One need not be Jewish or Muslim or Buddhist or Hindu or Christian to understand the clergy could not survive without the faithful.

Therefore, I expect the most aggressive antagonists against PhiAlpha will be the clergy, the clerics, the priests, and the rabbis. It is, after all, their geese that PhiAlpha cooks. It is their careers and pensions that PhiAlpha imperils.

PhiAlpha ultimately stands for the independence of our human existence from any religion. With PhiAlpha, religions have lost their power and control over whether we exist after our body dies. It is not that God is dead—but that religion is. Religion is now irrelevant to the issue of an afterlife and whether we "earn it" by either our faith or good works. They aren't relevant any longer.

PhiAlpha reveals what the priests and clerics knew all along but refused to tell us: the only motivation for decency and morality in the world is what we, as individuals, resolve on our own to follow. In other words, the Christian faith will not automatically turn a barbarian into a law-abiding citizen, or vice versa. Living together in peace is a choice one makes regardless of religious affiliation. It is not a matter of faith that we seek peace but common benefit. We each survive. Civilization depends more upon the unspoken agreements between individuals than it does our collective religious faiths.

15. DYING DAY

"What's wrong with you? Are you sick or something?" the girl with pretty blonde curls whispered to me, a frown on her face as though she was afraid of getting contaminated by my germs. "Your fingers are blue, and you sound like you are leaking. Gross."

Her name is long forgotten. I do remember she sat next to me in our first-grade classroom at Lafayette Elementary School in Phoenix, Arizona. A very nosey young girl, she was always asking me questions about myself. Being very shy at 6 years of age, I tried to ignore her as much as possible.

But on this particular day, my lack of response had nothing to do with being shy.

I couldn't speak. An invisible elephant was sitting on my chest, suffocating me, and I would have to somehow slay it to survive. My 6-year-old mind wandered through an oxygen-deprived haze.

Outside the classroom windows, the sky turned reddish-brown from the roiling sandstorm that had engulfed southern Arizona that day. My parents had made me walk to school, despite warnings that the sandstorm would have serious health effects, especially for kids, like me.

Staring down at my chest, I imagined my invisible beast sitting on top of me. Did the beast care that he was killing me by inches every time he sat on my chest? Would he feel guilty if I died?

Did he understand the difference between compassion and cruelty?

Maybe, if I showed him kindness, he would let me go. I tried to think of how to show my beast a kindness profound enough for him to allow me to live. Maybe, I should give him a name. If we exchanged names, maybe he would care more about me. Maybe I would no longer be the anonymous child he had chosen to kill. Maybe, if he knew my name, he would let me live.

"Beast," I offered. "Do you know my name? I am Michael. If I give you a name will you let me live?"

There was only silence. Apparently, the beast did not care to know my name. The beast confronted me with the ultimate cruelty—total indifference. It did not care who I was or whether I lived or died.

But I knew its name. I knew it well. Its name was "Asthma," and it had been trying very hard to kill me for all five years of my life. By the time I made it to first grade, I had been to the hospital twice with life-threatening bouts of double pneumonia. I was so sickly, that my mother kept me in a special bed every day. It was covered with a sheet. A steaming pot of water was placed at the side of the bed, under the sheet, supposedly to help me breathe.

It did not work.

My family finally moved me from Ohio to Phoenix hoping the dry desert air would help lessen the asthma attacks. This did not work either. What deserts lack in moisture, they make up for in those little granules called sand (aka dust). Around Phoenix, the blowing sand is everywhere—in your clothes, in your hair, and in your lungs. You cannot breathe sand.

That day in school was the worst asthma attack I had ever had. With every breath, my "elephant" sat on me harder. Each succeeding inhalation required more effort, more strain than the last—my chest caving inward like a punctured balloon. Each exhalation wheezed out of my lungs like steam from a leaking pipe.

The girl no longer existed. By the time she saw my fingers turn blue, all I could do was sit in my seat and concentrate on getting the next lungful of oxygen. Trying to breathe was like sucking air through a blocked straw. The more I inhaled, the less useful oxygen I seemed to get.

The teacher and the classroom no longer existed. My world shrunk down to the fundamental mechanics of breathing in and breathing out. Each breath purchased at the cost of complete mental focus, blocking out everything else. Each breath paid for with every muscle and sinew I could bring to bear to hold off my "elephant" for one more breath.

There was no other option. It was the ultimate, visceral high. Slay the elephant to breathe—or die.

The little girl poked me in the arm.

"Well, I can see you aren't going to do anything about it," she whispered with a disapproving shake of her blonde curls. "I'm telling the teacher."

The next thing I knew, I was in the family car, sitting in my mom's lap in the back seat, my head cradled in her arm. I raised my head. Up front, my grandpa was driving, my grandma to his right in the passenger seat.

Speeding away from the school, Grandpa drove through the red stoplight at the corner. Tires screeching, he turned onto a divided highway. The car accelerated rapidly. Grandma grabbed his arm.

"Paul, you should slow down," Grandma said.

The car went faster. I saw his reflection in the rearview mirror. I saw something I had never seen from my grandfather in my lifetime: his face was a grim mask of stone.

"How is he?" Grandpa's voice was tense.

Mom did not speak; she sat rigid in the back seat, crying softly, her fingers massaging my chest absently.

The car lurched forward, accelerating faster. Intersections sped past us like the film slides from a motion picture. But what was the hurry? I was fine now.

My inner self felt light as a feather—so light, I thought I could easily float away. There was no longer any pain. The elephant that had been sitting on my chest had disappeared. Had I slain my beast?

There was a subtle change in the air density just outside the car. The air outside the car quivered for a microsecond. From out of the distortion of air appeared a disembodied face. Ghostly white, it was. It floated toward the car from the outside—keeping pace with us. It hung motionless on the other side of the window next to me in the back of the car.

"Child, do you want to come with me?" Her voice was as gentle as a mother's caress.

The disembodied head was that of a woman. She was transparent. I could see through her to the sidewalk and buildings behind her.

No description here will be worthy of her appearance, but I will try. She was simply ... otherworldly. That is the only word that fits how she appeared. Long white hair framed a delicate yet mature female face. The face appeared to be an adult woman in her mid-30s but with a gravity surrounding her eyes that belied any superficial appearance of youth.

"Who are you?" I asked. She answered with a rush of syllables I barely could make out and then made me promise never to repeat it. I couldn't if I tried. I can give you a hint, though. Her name begins with an "Arie" sounding syllable. That is all I could make out before the remaining syllables followed in a torrent of sound.

"Do you wish to come with me?" She repeated, smiling compassionately. "I am your mentor. I am your one and only true friend in this realm. You do not have to struggle to breathe anymore. We can give you a better body. You deserve a better body. You deserve a body that is healthier and stronger. We can see that you are given one. All you have to do is tell me you wish to come with me. Just reach out, take my hand, and you will be free of all the needless pain and suffering."

Smiling lovingly, a hand reached through the glass of the car window as if the glass were not there. The hand stopped several inches from me and stayed there, open and inviting. The hand never tried to touch mine, never grabbed it. It was an offer, not a command.

"Just take my hand, child," she whispered softly. "Take my hand, and we can go together."

"Do you see her, Momma?" I whispered. "There is an angel floating outside the car. She wants me to go with her. She wants to give me a better body. Look! Look! Do you see her?"

Oddly, my chest no longer hurt. I no longer felt the need to breathe. Serenity and calmness enveloped me like a cloak. My body no longer felt heavy. If I tried, I felt like I could float away.

"He's not breathing!" Mom screamed.

"What?" Grandpa asked from up front.

"Poppa, hurry. He stopped breathing."

I watched placidly from the back seat as Grandpa sped toward the hospital. Telephone poles flew by like twigs in a stormy wind as we raced along the divided highway. There is an elegant purity in feeling so weak you do not care anymore whether you live or die. You realize you are in the hands of a higher power, like a puppy in the hands of its new owner. Complete helplessness has its own freedom.

When we arrived at the hospital, Momma scooped my limp body up and rushed me through the double doors into the ER, but there was no one there. The ER was empty. There were no other patients. The only medical staff was a young physician and two or three nurses. The doctor was flirting with one of them. They were taken completely by surprise. One moment they were joking around, bored out of their minds; the next moment, they turned around startled by our sudden entrance.

Then they saw me. One of the nurses was the first to react. She quickly took me from Momma and held my body up for the doctor to examine.

"He's going blue."

The ER doctor recoiled away from me. I sensed panic in him. I felt his self-doubt, his questioning of his own competence. Could he save me? How could he save me? What would happen if he couldn't? Was his career over before it had really started?

The ER doctor had come to work that day with no medical backup. He had not arranged to have anyone on call in case of emergency. He was on his own. He had not been expecting anything more serious than a cold to come in—or at worse a broken arm. He had not anticipated life-or-death today. But, that is what he got.

The nurse laid my body on the ER treatment table and stripped me down to my underwear. Four surgical lamps, arranged in a circle, were positioned above the treatment table for illumination. Underneath their bright lights, my limp body was turning bluer by the minute.

Part of me separated from my physical body and began floating upward toward the ceiling. This astral self was connected to my body by an umbilical-like energy cord that hummed as it vibrated. It curled beneath me like a snake. This astral self rose upward of its own volition.

I had no control over this astral self, nor could I see it. I could see my physical body, lying on the surgical table below. But whatever I had transitioned into, I could not see it. I do not know what my astral self looked like.

Whatever it was, it had limitations. While I could "see" and "hear" all that I focused upon in the ER, I had no sense of smell or touch. Nor could I speak to any of the humans below.

My astral self floated upward, above the treatment table. As I reached the surgical lamps, I got anxious. Where would I stop? Would I float right through the ceiling? Or would I just disappear into nothingness? This was all too strange for me to process.

Fortunately, my astral self stopped rising halfway between the surgical lamps and the ceiling. It felt like I was tied to a helium balloon. There was a small but perceptible up-and-down and side-to-side buffeting. There was no visible cause for this lazy motion, but the motion was distinct.

I had no idea what was causing it at the time. But, looking back, the answer is obvious. Hot air rises. The air vents in the ER were cycling fresh air in all the time. This air current was enough to buffet my astral body.

The astral body created its own problems for me. I could not control where it went. I tried to think it lower and higher, to the right and to the left. I used my imagination to move whatever invisible arms and legs my astral self might have to "swim" up and down, right to left. Frustratingly, nothing worked. I was helpless up there, above the lamps. My physical body was dying below, and there was nothing I could do but watch.

But even watching was difficult. For some unknown reason, my astral self seemed to be naturally attracted to the four surgical lamps. I was desperate for an uninterrupted view of what was happening to my physical body below, but my astral self refused to move away from the surgical lamps above the table. There was enough slack in the umbilical cord to allow movement far enough away for a clear, unobstructed view, but my astral self was operating on a separate volition than mine.

Totally detached from all physical sensation weirded me out. I needed to connect with my body on the treatment table. But where does one look within one's mind for physical sensation? I tried to sense what the treatment table must feel like, but there was no feeling whatsoever.

The young doctor pushed his self-doubts aside and let his training take over. He turned to Grandpa.

"How long has it been since he stopped breathing?"

"Just as we entered the ER." Grandpa replied.

"Doctor, his heart just stopped," the nurse reported, tensely.

From just above the four lamps, I watched the doctor treat me. He repeatedly compressed my chest. Every so often he breathed into my mouth,

"Are you coming with us?" A male voice spoke to me from above.

I looked around. A dozen or so oval-shaped orbs of light had suddenly appeared next to me above the treatment table lamps. They did not have any distinguishing features, just blank orbs of light. But they all carried with them personalities. I could "feel" their different identities. They were individual presences—and they were in a hurry.

More of these orbs were appearing from behind them. Their entry was crowded but orderly. Where were they coming from?

Then I saw it. Near the ceiling was a silent but intense circular pulsation of light and shadow.

What did this object look like? Let me give as objective a reconstruction as possible.

Increase the chaotic intensity of a tornado's interior at least 10 times, turn it on its side, slice it in half vertically, and there it is—the portal I saw near the ceiling.

A dozen or two of the pale white orbs came and went into it and through it like we walk through a door. There was no drama about their entry or exit, no excess effort. They did not hesitate before going through it, and their passage occasioned no additional effects upon its appearance. Though it was unsaid, I understood that this portal was my way out as well.

The spirit guides emerging through the portal hung around near the ceiling, apparently watching the emergency medical procedures below but keeping a respectful distance. Their pale white orbs were moving from side to side and up and down in the air, just like I was.

My ability to see the walls and ceiling of the hospital was affected by their presence. Where they stood, there was no hospital ceiling, or wall. Their entry into, and continued presence within, the ER had somehow extended their realm into ours. Whatever rules of matter and energy existed for them obviously took precedence over our own. Or was it the same rules but just different wavelengths, focus? Or was it just different matter and energy—like dark matter and dark energy?

I had no clue then nor any better clue today.

One of the orbs approached my astral self above the lamps. As it came nearer, a regal-looking male face appeared on its surface facing me. I knew instinctively, this was the face behind the male voice that had just spoken to me.

Was the face real or manufactured? The face appeared more like a mask than a real face. Though it had facial features, there were no wrinkles or lines that real faces have. My conclusion was that it was manufactured just for my benefit.

"You need to hurry," The male face in the orb ordered. "Time is running out. The young doctor has been presented a puzzle he may not solve in time to save your life. The brain only has so much time to survive without oxygen. You must decide now. If you do not choose, we will have no choice but to take you. We are not waiting for him."

Was he their leader? There was a strong "aura" of authority around and about him. He obviously expected to be obeyed.

Was this ego? They were operating under what appeared to be a fixed time constraint. So, his behavior was more likely the result of an intrinsic operational command authority than ego.

What stood out was their objectivity. Their objectivity varied from presence to presence among them. I had expected a sameness about them, a lack of personality, but as they came near me, I could "feel" their individual personalities in the same way humans sense fear or excitement in others.

I felt a very strong emotional connection with the female spirit guide I had met in the car. She projected feelings of love and compassion for me that made me want to go with her. By comparison, the male "leader" was stone cold. All I could feel from him was his pride in his objectivity. He wore it like a crown, as though it was his proudest achievement, something his peers were training for—yet lacked.

Allow me to confess something. I never did tell the female spirit guide whether I wanted to leave or not. That was on purpose. I was only 6-years-old, for crying out loud. Six-year-olds aren't supposed to make such do-or-die decisions, are they? To hear her talk, though, perhaps they are.

Anyway, I was afraid any decision I made I would regret later, so I avoided her question. Our arrival at the hospital had interrupted the matter. Now, the spirit guides attending me above the lamps were at me again.

"Decide now," the male spirit guide commanded.

He intimidated me. The energy of his thought reverberated through my psyche like a lightning rod. I felt like a drill sergeant of a celestial boot camp had just told me to drop and give him a hundred spiritual pushups. Somehow, he was able to focus his thoughts into mine not only to convey a message but the emotional force behind it as well.

They were right. The body I had was pitifully frail. I deserved better. Below, the poor doctor was becoming increasingly desperate as his efforts to get my heart restarted were unsuccessful. This was not going to end well. The umbilical cord attaching me to my body was growing dimmer and dimmer.

It was time to say goodbye.

I paused to look at my parents. They did see not see me. They had not once looked up in my direction, above the lamps. All they saw was their child on a surgical table, dying. But I was not there. I was floating 10 feet above them.

What I saw and felt broke my heart. As families go, we were pretty closely knit. I had expected the tragedy of my death to bring my mother and grandparents together. To my shock and distress, I could not have been more mistaken.

They were not huddling together, coming together in their grief. Instead, they had withdrawn from each other. Each trapped within a private prison of grief.

Mother, alone by herself, whimpered into her lace handkerchief, staring into space by the wall—as far away from my body on the treating table as possible. Grandma stood expressionless halfway between Mother and Grandpa. Her Native American heritage had given her a stoicism I could not read.

Grandpa stared blankly at the double doors we had just come through. Hands thrust deeply into the pockets of his gray pinstripe suit. His face pale, arms visibly tense inside his suit coat—a man who had lost his soul.

My poor Grandpa.

Grandpa was one of those guys others would call a self-made man. I worshiped him—and I was not the only one. Almost everyone in town knew my Grandpa. He treated everyone the same—from the town mayor and police chief to the trash collector and neighborhood chimney sweep.

Every day at lunchtime, he would walk me uptown to the courthouse square. Placing me up on the courthouse steps, he would introduce me to all the people he knew—and he knew almost everyone that walked by. For a preschooler, that was a big deal. He made me feel important, necessary.

Down below me, however, the Grandpa I knew no longer existed. He looked lost and forlorn standing there, staring at those double doors in the ER without seeing anything at all.

"Last chance, child." The female spirit had returned to my side, floating next to me above the lamps. "You can stay or you can come with us, and we will give you a better, healthier body to play with." She was smiling, yet pushy—like a parent hurrying a tardy child to get dressed.

Behind her, I could see all the other spirits disappearing into the portal. The male leader must have gone with them. Their exit was as orderly as their entrance had been.

Why had they appeared at all? What was I to any of them? Did they regard my medical crisis as one would visit the birth of a kangaroo? Was I an exhibit at some celestial zoo?

Below, like my mom and grandma, my grandpa remained frozen—staring blindly into space. Did he feel he was to blame for what happened to me? If so, it would haunt him the rest of his life. The guilt of tragedy without warning is the hardest guilt to overcome.

If I left, I would doom them to spending the rest of their lives trying to explain to themselves and each other why, asking themselves, "Could I have done something to prevent my child's death?"

If I left them now, they would no longer be the close-knit family I had known. For the rest of their lives, individually and collectively, they would be trapped in this moment of despair and self-recrimination. That would condemn each of them to a living hell.

I turned toward the female spirit guide.

"I'm staying."

"Are you certain, my child?" She asked.

"I will make this body better on my own. You wait and see."

She gave me an odd glance; one-half a question, the other half an unspoken judgment. Would she overrule me? Could she overrule me? Time stood still for me.

Then, with a smile, she nodded.

"Nurse, 10 CCs of epinephrine, stat!" The young doctor appeared to have had a sudden inspiration. "Why didn't I think of that sooner? Dammit!"

The doctor had gone from groping and floundering without success to sudden decision and action. The change was dramatic. Apparently, he had experienced an epiphany of some kind. Had the spirits mentally suggested to him what needed to be done? After all this delay, would it be in time?

The nurse handed the doctor a hypodermic syringe with a long, thick needle. The doctor did not wait. With a quick thrust, he plunged the needle into my chest and pushed the syringe down, injecting the epinephrine directly into my heart. I felt nothing. I was still floating above the surgical lamps.

As soon as he withdrew the needle from my chest, the lead nurse breathed into my mouth and pushed my chest down rapidly.

Without warning, I was back in my body. The process was instantaneous. One moment I was floating above the lamps in the ER. The next moment I was back in my own body. The steel table was ice cold on my back. I squirmed on top of it.

I chanced a deep breath. I inhaled as deeply as I could. To my great relief, my lungs felt completely free of any asthmatic constraint. I enjoyed breathing normally so much I got goofy. I was high from the epinephrine. I bounced up and down on the treatment table.

"Stop it," barked the nurse. "You'll break the table."

But the young doctor was on my side.

"Nah, let him bounce," he said, laughing. "After what we've been through, I feel like bouncing myself."

16. CAMELBACK MOUNTAIN

I spent the next two weeks recuperating at home. Our home was a one-story ranch on East Pinchot Avenue. Back in 1956, our house was on the very last street in northeast Phoenix. Beyond our backyard was just desert. And beyond the desert was the mountain—Camelback Mountain.

One morning, within a few days of my return home, I was resting in bed when I heard a familiar female voice.

"Feeling better are you now?"

My female spirit guide had returned.

"Why don't you get up?" she prodded. "If you want to get healthy you can't play the sick child anymore. You have to start pushing yourself."

"What should I do?"

"Start slow," she suggested. "When you are older and stronger, you can exercise and play sports. For now, why don't you get up, go outside, and see what your grandma is doing?"

That sounded easy enough. I got dressed and walked through the kitchen and out to our backyard. Grandma was at the back fence hanging laundry on the clothesline, staring off into space as she did so. A bag of clothespins hung from the line. On the sandy ground next to her was a basket full of freshly hand-washed clothes.

Stooping down, Grandma picked up a white blouse out of the basket and tried to hang it on the line. Two clothespins fell from her hand to the ground.

"Can I help?" I asked. I picked the clothespins up and gave them to her.

It was then that I first saw it. It was so far away, the brown-and-tan complexion of its western side shimmered in blue and violet shadows in the morning sun, yet the mountain was still so intimidatingly huge I wanted to run back inside. Because our backyard was at the extreme edge of east Phoenix, there was nothing between our back fence and this raw, barren desert.

"Golly, what is that?" I asked as I pointed.

"That is Camelback Mountain, Michael," she answered, staring straight ahead toward the mountain. Her voice was distant, trancelike.

I realized that must be why she dropped the clothespins. The clothesline was strung across our backyard so she could hang the laundry without taking her eyes off the mountain. Her attention was focused more on the mountain than the clothes she was hanging.

"Why is it called that, Grandma?"

"Look at the mountain, Michael." Her voice became her own again. The odd evenness was gone. "See the two humps? The white men called it 'Camelback' because they said the humps look like the head and hump of a camel resting. But for the Native American tribes that were here first, the mountain is sacred. Native Americans do not see a camel. We see the head and belly of a pregnant woman. That is what we call it, 'The Pregnant Woman.'"

Pondering what she had said, I stared at the mountain as though it were a sacred artifact of a brave and holy people, now long lost. Frustratingly, despite my best efforts, I was unable to see any camel—or pregnant woman.

"Grandma, if they weren't here first, why did the white men get to name the mountain?"

Grandma stopped hanging up the blouse and looked at me. It was the first time she had glanced at me since I came outside.

"You should go back inside, Michael," she instructed. "You must be getting tired."

Apparently, in 1956, some things were best left unsaid.

"But the angel woman told me I need to push myself," I countered.

Grandma's head turned to the side, and she took a moment before answering.

"Yes, but I don't think she would want you to push yourself too much too soon, now would she?" She said seriously.

I had to nod. Yes, that was true.

"You have the rest of your childhood to push yourself, you know," Grandma said wisely.

I was OK with that. But before I went back inside, I had to take one more look at the mountain. Turning, I leaned up against the fence and fixed my focus one last time upon the strange shadow looming imposingly out of the northeast.

Suddenly, I was out of my body. My astral self flew out over the brush, above the cactus plants, and across the desert sands toward the mountain. As I streaked onward, the mountain appeared to rise up out of the sand at me like a lion leaping toward its prey.

The vision ended as quickly as it had started. In a split second, my mind was back in my body by the fence in my backyard. I stood there wanting to rush out into the desert right then and there.

"Not now, my child," spoke the female spirit guide. "Your time will come soon."

I was hooked. I made a vow then and there. One way or another, I would get to go to Camelback Mountain. It was my sacred mountain now, too.

Was it a vision? Or was it simply the hallucination of a young boy still recovering from a near-death experience? Six-year-olds do not care about causation or the terminology to rationalize or euphemize life. I was still too young to place experiences into linguistic boxes created by school education and socialization. I was just living it.

Planning for my journey to Camelback Mountain was as simple and direct as a 6-year-old can make it. The mountain did not look to be that far away. I figured it would take two to three hours to walk there and back. If I skipped out of school right after the 2:15 p.m. recess, I could get home, pack a lunch, and be on my way and back home before dark.

No problem.

But what if I chickened out? That would not do. I would not be able to live with myself if I chickened out on going to my mountain. I would have to come up with some strategy to keep myself from changing my mind at the last minute and chickening out.

My first day back at Lafayette Elementary, I made sure to say hi to my buddies. They were all happy to see me. But there was one classmate I was especially interested in talking to—my best friend, Mark. I had a favor to ask him—a big favor.

Mark was in another class, so I had to wait for recess to meet up with him. When recess finally came, I rushed out to the playground. Sure enough, there he was, playing basketball. The school had a nonregulation basketball court. Regulation basketball courts have a 10-foot-high hoop. Our school's hoops were only eight feet high but still too high for me to reach. I did not have the strength to shoot the ball that high—even two handed.

"Hiya, Mike." He smiled and bounced the ball to me. "See what you can do. Take a shot."

I caught the ball and took a couple of awkward dribbles. Some older boys were at the other end of the court had started watching me. A couple of them were already pointing in my direction, laughing.

"Nah, here." I tossed the ball back to him. "You play. But I have a favor to ask."

"What's that?" Mark asked.

I told him what happened at the hospital and about my plan to go to Camelback Mountain.

"You're crazy," he said with a snort. Mark shook his head angrily. "You want me to play hooky from school and get my ass in trouble just so you can go to some stupid mountain? You … are … nuts! Do you know what happens to kids that get caught skipping out of school?"

"No," I confessed.

"They get suspended, that's what." He laughed at me, but he wasn't smiling. "Then, your parents get ahold of you. Do you know what happens when my dad gets mad at me?"

"No."

"He beats me. That's what he does. Don't tell anybody or I'll kick your ass." Mark rolled up the left arm's shirt sleeve. There was a strawberry-red welt on his arm, just below the shoulder. Mark pulled the shirt back over the bruise. "I ain't getting beat just so you can go to a stupid mountain."

Mark started to walk away.

"Fine," I replied. "It's better that you don't come along. But I'm going. I have to go."

"You go, you'll either end up getting bit by a rattlesnake or eaten by a panther," Mark said, shaking his head. "Either way, you'll end up deader than my old man's car."

"I should already be dead. I'm still going." I turned and began to walk away.

"Wait."

I turned around to face him again. He had picked up one of the basketballs.

"Tell you what, Mikey, ol' buddy," Mark said smiling. He was spinning the basketball in his hand as he spoke. "See that basket over there?" He pointed to the nearest basketball rim about 15 feet away.

I nodded.

"You get one shot and one shot only. From the free throw line. You make that shot, I'll come along with you to your stupid mountain."

"Really?"

"But if you miss, I don't go and neither do you. What do you say?" He grinned mischievously. "Deal?"

"Deal." I strode over to the free throw line with all the bravado I could muster.

Kids who had been playing ball at the other end of the court started making their way down to our end.

"What's going on?" one of them asked.

"What's up?" A second kid wanted to know, too. In short order, the half-court we were at had filled up with well over 20 kids.

"Hey, Mikey here is going to try to get me to go with him to his stupid mountain. If he makes the basket, I go. If he misses, I stay and so does he." Mark laughed. "He's never made a basket in his life, I think."

That made everybody laugh—except me.

Mark tossed the ball to me and pointed toward the free throw line.

I was pissed off now. Mark had made sure he embarrassed me in front of his buddies. Screw him.

In the past, every time I had tried to shoot, even from in close, I could barely reach the rim. There was no way I was going to make a basket from way out at the free throw line. I dribbled a couple of times and took aim. I balanced the ball in my right hand and threw it up at the basket without thinking.

Swish! Nothing but net. I couldn't believe it. The kids yelled, clapped, and cheered—all except for Mark. Mark glumly stood by himself, staring at me. He was rubbing his left arm.

"Promise me you won't get us into trouble," he demanded.

"Don't worry," I replied, patting him on the back. "We'll be fine."

"You better be right," he said. "Cause if I get beat, I will totally destroy you."

I decided we would go the very next day. Even though I had someone to go with me, I was afraid if I put off going too long, my compulsion to do this would fade away. Even then, I knew there was something odd about my sudden urge to go to the mountain—but I was feeling too "in the zone" to think straight, I suppose.

Now I had a much more immediate issue at hand. I didn't want anyone to know about my hike to the mountain. Including Mark in my plans was a necessary risk. I needed him to go with me. But now, after the basketball court showdown, half the school probably knew about the trip. Who knew how many kids would rat us out? There is usually at least one punk who always tattles on stuff like this.

I didn't sleep that night. My thoughts jumped from the exhilarating expectation of hiking to Camelback Mountain to sheer terror at the thought one of the kids at school ratting me out. It was frustrating. Every happy image of hiking over the desert to my mountain kept getting swallowed up by phantasms in which boys from the basketball court were telling their parents about our hike, and the parents were calling the teachers, and the teachers were calling our parents—and poor Mark getting beaten by his dad. Waiting became torture. I feared the worst.

But the day would not be denied, and I trudged to school regretting ever getting anyone else involved. This was my one and only chance to go to my mountain. If anything happened to prevent it, I didn't know what I would do.

When I did get to school the next morning, I was shocked. To my great surprise, the school day went like every other school day. No one said anything to me about skipping out of school or the planned hike. Anticipation built as the day went on. By lunch, it was all I could do not to shout at the top of my lungs what I had planned to do that day.

As soon as 2:15 p.m. recess came, Mark and I were out the school door silently and quickly and headed to my house two blocks away.

Once we got to my house, Mark stayed outside. I went inside to the kitchen. No one was around. Grabbing a paper bag, I filled it with two cans of bean soup and two sodas. I also scooped up two spoons, a can, and bottle cap opener. In seconds, I was out of the house with our loot. Mark and I scooted out of my backyard without my folks knowing we were ever there.

"Hey, we did it!" I clapped Mark on the back as we started walking. We had made it. We were in the desert. We were on our way to my mountain. We'd made it 15 yards into the desert.

Mark grabbed my arm. We both stopped.

"What?"

Mark raised his arm and pointed ahead. Out in the distance, far enough away that the air rippled in the heat, loomed a giant shadow. It rose up out of the desert floor like a sleeping giant. That was my mountain, Camelback Mountain.

"How far is your mountain from here?" Mark quizzed me.

"I don't know," I replied. "We can make it there and back before dinner, for sure."

"Look, the only reason I am on this little hike of yours is to make sure you don't get yourself killed. Just don't get me in trouble," he warned. "I don't wanna get beat." He grabbed me by the arm again. "Remember what's gonna happen to you if I get beat."

"Right," I replied. "I get destroyed. Look, there's nothing to worry about," I cajoled, patting him on the back. "We're going on a great adventure. Think about it. How many first-graders hike to a mountain on their own? We're making history."

Can't be too far, I hoped.

Back in 1956, what is now Paradise Valley was just desert behind my house. Camelback Mountain's holy ground was yet unblemished by hiking trails, golfing resorts, and spas.

Camelback East Village did not yet exist. For that matter, neither did Pasadena, or Tierra De Paz, or Camelback Heights, or Red Rock North. There was no Phoenix Country Day School, Arizona Canal Trail, Phoenix Swim Club, or Kachina Park.

People had not yet driven down Palomino Road, or Calle Del Norte, or Montecito Avenue, or Roma Avenue because these roads had not been paved yet. The entire area northeast of Phoenix all the way to Camelback was nothing but desert, miles and miles of desert.

But back then, both Mark and I both thought Camelback was nearby. It must be to look so big.

"Come on, let's run," I yelled and started to run.

"Wait!" Mark wasn't running. "It's really hot."

Mark was right. I realized I was already sweating profusely. This was not good. We had not even gotten out of eyeshot of my back yard and already I was dying in the heat.

"Oh, shit."

It occurred to me that maybe walking in the desert in the midday heat was not such a great idea. I obviously failed to take that into account in my planning for the trek.

Walking into the desert in the middle of the afternoon was like walking into an open furnace. It may have been the end of September, but it was still summer hot. The sun beat me down from above. My uncovered head of hair felt like it was about to catch fire. Heat rose up off the sun-seared desert floor in waves, burning my feet through my sneakers. Just standing around, we were sweating through our T-shirts.

I did a quick mental computation. If we stood around any longer, we would boil. If we went back, we failed. There was only one option.

"If we don't run, we risk getting caught," I urged Mark on. "We need to run until we know they can't find us. Come on, we don't have to run far. Once we are out of sight, we can walk."

I took off running into the roiling desert heat. Eventually, Mark followed.

We ran past cactus plants and brush; we jumped over rattlesnakes and tarantulas; we waved at gophers and coyotes. We both learned to use the sights and sounds of the desert to take our minds off the will-sapping desert heat. It was mind over matter—or die. But more than anything else … it was a blast.

Mark laughed more than I had ever seen him laugh before. It made me happy to see he was so happy. We were on a grand adventure and we were only *6 years old*. I had never been so happy in my entire life. Fate had found me. I was where I needed to be at the time I needed to be there.

"Imagine," I yelled to him. "No one does this at our age. If we can do this adventure at age 6, imagine what we can do when we are 26."

Due to the relentless heat, we had to alternate how we traveled. When we couldn't run anymore, we walked. The farther we went, the more we found ourselves walking. Mostly, we walked. The walking never ended ... and Camelback Mountain was still no closer.

I let the monotonous routine of walking take over my mind. I lost awareness of where I was and what I was doing. I had found the space between time and matter. I got high off it.

But there was one positive: I breathed clear and deep. There was no sign of asthma. I couldn't believe it. For the first time in my life, I was running and running free of chest pain. The elephant was gone. The beast had been slain.

There was one problem, however. No matter how far we walked, the mountain never seemed to get any closer. After an hour or so, Mark became frustrated. He repeatedly complained about how long our hike was taking. Doubt was creeping into his mind about whether we would ever make it.

The heat quickly became too much for us. We stopped to rest in the shade of a huge boulder. I pulled the soup and soda out of the bag and handed one of each to Mark. Leaning up against the boulder gave us needed respite from the heat. Eating our rations in the wild was a special moment for me.

"Let's go, let's go ..." Mark was apparently less in awe of the moment than I. We left our empty soda bottles and soup cans by the boulder and ran on toward the mountain.

The running gave way to walking fast. The walking fast soon gave way to walking slowly. Always, Mark lingered behind me. He was my body guard, but there was really never any threat to us beyond the desert heat. We saw no panthers or mountain lions. The occasional rattlesnakes we came upon were too lazy sunning themselves to be any threat.

"Hey, stop," Mark called from the rear.

I stopped walking and waited for him to catch up.

"Look." Mark pointed back the way we came.

There was nothing to see except desert. Tiny white lights flickered off and on in the far distance.

"I can't see anything except little lights." I reported.

"That's the point, nutball." Mark was angry.

I didn't understand why.

"Know what those 'little lights' are?" His voice was agitated.

"No."

"Our neighborhood, you ass-wipe." He said as he backhanded me on the arm.

"We walked so far, we can barely see home, and we're still not at your stupid-ass mountain. Look at the sun! Shit, just look around!"

"It's setting." I answered. The sun was about to set. I guessed it was 5:30-6 p.m.

"That's why the lights are on, dimwit. It's already dinner time." Mark shook his head.

"You know, the only reason I came along was because I promised to protect your ass out here."

"I know. I appreciate it."

"You'd better," Mark said with a heavy sigh. "Because I'm gonna get my ass beat and then ..."

"I know, I know." I already knew the drill. "You get destroyed."

Mark raised his arm.

I closed my eyes and braced for another impact.

Instead, Mark put his arm around my shoulder.

"Ya know, I thought you were kidding." Mark wasn't mad. "I didn't think you'd go through

with it. We all thought you were a dweeb, but you got guts. It was a lot of fun, but it's getting late, and I need to be back before dark. You should come with me."

We stood side by side in the desert. The cactus plants and boulders surrounding us were growing shadows in the dwindling sunlight. I felt a shiver. The air was causing goosebumps on my bare arms. Who knew? One minute I am near heat stroke, and the next I get cold. The desert is full of surprises.

"I'm not going back." I told him. "Come with me."

Mark wouldn't look at me.

"I can't." His voice quivered. "It will be dark soon, and we won't be able to see. There's no point to this anymore. If you keep going, you'll end up dead. You know that, right? Let's go back now." Mark wiped something from his eye.

"I can't," I said, shaking my head. "I came this far, I'm not going back until I reach my mountain." I patted him on the shoulder. "Tell your dad this was all my fault. It was my idea. You went along just to protect me. Tell him he can beat me—but not you."

"Forget about it. Shit, I get beat every day anyway. It doesn't matter. But this…?"

Mark held out his right hand. Two kids, alone in the desert, shook hands. "This matters. See ya, buddy."

Mark waved as he started off on his way back to Phoenix. I watched him disappear among the shadows. Maybe I should have gone back with him. Would I ever see him again? A feeling came over me that this was the last I would see of him.

A chill ran up my arm. The sun was setting. There was very little daylight left to make it through the desert. During the day, I had no fear of encountering desert predators such as coyotes, mountain lions, cougars, wolves, and rattlesnakes. They were night stalkers. I knew the heat of the day kept most of them in their burrows and caves. But as the darkness fell around me, I had to fight down a growing blind urge to run after Mark as fast as I could. Looking around me at the growing shadows, a chill ran up my spine that had nothing to do with the cooling desert air. I turned and began walking as fast as I could toward my mountain.

The shadows of sunset were becoming a hazard for me. I could not see very far ahead, and every so often I tripped over … whatever it was that I could not see. Fortunately, I was lucky enough not to break any bones. But I got beat up.

"Do not run," I said to myself. "Do not run."

That phrase had become my mantra walking ever deeper into the desert toward the looming mountain ahead. Only two animals ran in the evening in the desert—predators and prey. I had no chance of being the first—and I had no chance of survival if I became the second.

"Mike," I said to myself. "You have two jobs to do if you want to survive. Be as quiet as you can, and do not fall and get hurt. You get hurt out here, no one will find you in time to save you. You'll be breakfast for the cougars for sure."

I walked as silently and as fast as I could.

Night fell upon me sooner than I expected. But with the night came an unexpected miracle, a full moon. I knew nothing of lunar cycles, so this was just blind luck. A cloudless nighttime sky allowed the moon's reflection to light up the desert around me. I saw better during the night than during sunset. At least I wouldn't trip over a rock or walk straight into a cactus.

Night also brought a new threat I had not thought of during my pre-hike planning. Sunset had taken away the desert's heat engine.

With nightfall, the temperature fell dramatically. I was only wearing a short-sleeved T-shirt and khaki shorts. I hadn't planned on a nighttime trek through the desert. I thought I would get to the mountain and back home before dinner.

Walking in the night was surreal. When it got dark enough to see the stars, I felt like I was walking on the moon or on Mars. Everything was alien. Every boulder, every cactus had its distinct outline of shadow and light.

I began to lose my focus. Instead of concentrating on the task of walking fast in a straight line, I began to wander from side to side as I walked. At first, I couldn't figure out why.

Then I realized the problem.

My leg and arm muscles were shivering. Rubbing them only gave me temporary relief. Within a minute, I was shivering again—but worse. Rubbing my legs was taking time that should be spent walking. But, it could not be helped. I couldn't walk in a straight line shivering like that.

Very quickly, the cold got too much for me to endure. It hurt. My skin felt like it was burning. I stopped to rub my arms and legs again.

I got desperate.

I could not get any colder and live. As soon as I stopped rubbing my legs I planned to run as fast as I could. I did not want to run in the desert in the dark, but if I was going to survive, I needed to get back some body heat. It wasn't the best plan, but I was in trouble, and I knew it.

I stopped rubbing my legs and started to run, but before I could get two steps, my entire body spasmed. My legs, arms, chest, and sides shivered uncontrollably.

Unable to control any of my muscles any longer, I fell to my knees. I tried to move my hands to rub my legs, but my body refused to cooperate. Uncontrollable shivering made me totally helpless.

I lay on the desert floor, curled up in a quivering ball of flesh and bone, and I waited to die.

The cold had won. It wasn't fair. I had survived the sun and the predators of the desert only to fall prey to something I had not even thought about—the freezing cold of night.

My mind seized. I could not think. The unrelenting cold was more than I could bear.

Death curled up beside me. The cold was replaced by its now-familiar neutrality. I felt no fear. Death was not a stranger to me. I had seen its face in the ER. I had felt its silky smoothness slide like a needle inside me once before.

"Pick up a handful of sand."

Was it a voice? Was it a thought?

It was both and it was neither.

Was I saved again? Could I be that lucky twice? I fought to speak.

"Hey, I'm freezing here. Give me a blanket or something. Please." My voice a weak quiver in the icy darkness.

"Put the coldness aside and pick up a handful of sand."

There was a bright flash of light. I was blinded by it, but only for a few seconds. The flash of light had done something magical.

The cold was gone. The uncontrollable shivering ceased. I was warm again.

"Pick up a handful of sand and stand up," the voice instructed.

I dug my tiny fingers into the sand. I was surprised to find out the top layer of desert sand was not soft like a backyard sandbox but extremely hard. Baked by the sun day after day, all the moisture had evaporated long ago. My fingers pushed through the hard desert crust, into the softer sand just below. I scooped out a handful of sand and held it tight in my fist.

"Here," I stood up, holding out my fistful of sand out in front of me.

"Open your hand and watch the sand falls through your fingers," the voice said.

Little fingers relaxed their grip. As my hand opened, sand began to slip through my fingers. In the moonlight, the grains of sand glowed like golden magic dust as they fell toward my feet.

"Look up at the sky."

Obeying, I looked up at the nighttime sky. My little fist squeezed tight. I could feel the sand still slipping between my fingers toward the desert floor.

"Do you see the stars in the sky?" the voice asked.

"Yes."

"You are seeing just a fraction of the universe from where you are," the voice said.

"There are more stars in the sky than the grains of sand you hold in your hand. And there are as many galaxies in the cosmos as there are grains of sand in this desert. And every galaxy has billions and billions of stars. Each star has planets and countless planets have beings on them—just like you. Yet, no matter how many stars there are, or how many planets, or how many beings, none are any more or less special than you."

A shiver rose up my spine. It was as though a cloud of joy had burst right over my head. Tears flowed down my face. I laughed and giggled and cried at the same time.

"Will you remember this?" the voice asked.

"Yes. Oh, yes." Tears were still flowing as I giggled hysterically in a fit of silliness. I was shaking again—whether from the excitement or the cold or both, I did not know.

"If you wish to live, pick up brush and whatever else you can find, and cover yourself with it for warmth. You will be found before dawn, but whether you are found alive or dead is up to you. Do you understand?"

"Yes."

I felt rather than saw the voice leave. The voice must have taken the joy along with it, because as soon as it left, I started to shiver. I was feeling very, very cold again—and very much alone.

Quickly grabbing what brush and other scrub laid nearby, I fashioned a makeshift mattress where I could protect myself some from the cold desert floor and laid down on it. It wasn't perfect, but it would have to do. I tried to place some extra brush on top of me, but it kept falling off.

Laying back on the handmade mattress of desert brush, I stared up at the night sky. Staring back at me were untold billions of stars. Some I could see; the vast remainder I could not.

Questions came that I could not answer. What beings had the voice been talking about? Where were they from? When would I meet one? Why did the voice mention them at all? Why me?

Suddenly, I understood why I was there. I sat up and laughed out loud. My laughter echoing loudly into the darkness.

Whether one called it Camelback or Pregnant Woman, it had never been my mountain to begin with. Sacred mountains cannot be possessed by mortals. Even if occupied by hiking trails, golfing resorts, or spas, sacred mountains are impervious to our transitory human existence.

The sacred mountains of Earth (Everest, Mount Kailash, Mount Fuji, Uluru, and hundreds of others) are eternal and treat us as the earthly children we are. They teach us the only life lesson that matters—with stillness—that there is an infinite emptiness existing between each thought connecting us with our inner selves and each other.

The stars above and the races of beings inhabiting their planets may or may not ever meet us. But for a child of six, it was enough to know that no matter how important they were, they were no more—or less—important than me.

My laughter echoed, again, into the night. Live or die, I was happy. I laid back down on the scrub brush and let sleep take me wherever it wanted.

I awoke bathed in a blinding white light.

Whuup, whuup, whuup, whuup…

I sat up straight.

Something was coming at me.

Whuup, whuup, whuup, whuup…

Whatever it was, it was loud—and getting louder. The hair rose on the back of my neck. This menacing sound was accompanied by a howling wind that grew stronger as the sound approached me.

Whuup, whuup, whuup, whuup…

There was a thud. From the sound, I could tell something huge had landed only about a hundred feet away from me, but with the white light blinding my eyes, I could not see anything.

The voice had mentioned space beings existed by the billions. Had some of them come to get me? Should I run—or fight? I stood up, just in case.

Whuup, whuup, whuup, whuup…

Two shadows, each much taller than I, emerged out of the blinding light as though they were stepping out of another universe into this one.

Must be aliens …

"Hello?" I shouted.

But the howling wind shattered my greeting and through it back at me. There was no response.

The two "aliens" reached me. Neither "alien" spoke a word to me. One wrapped me in a blanket and the other scooped me up and began running the way they had come.

As they ran back to their craft, one of the beings made a gesture across his throat. Were they going to kill me? I kicked my legs, trying to get free.

"Dammit, kid! Stop kicking." The "alien" holding me just held me more tightly.

Suddenly, the blinding light was gone. I was able to see the "beast" that had landed. It was not from outer space. It was just a U.S. Army helicopter.

The two beings who had grabbed me were not aliens from another planet. Instead of space suits, they wore U.S. Army aviator flight suits. They were part of the helicopter's crew.

The crewmen hustled me aboard through the side hatch. The helicopter had two floors—one down below where the crew and I were, and the flight deck above. Connecting the two floors was a small set of steps.

I ran up the steps.

"Hey, kid! Stop!" one of the crewmen yelled from below.

"Hi!" I yelled. Two faces turned as one toward me—one from the right seat and one from the left.

"You OK, kid?" the co-pilot on the right asked with a friendly smile. "You gave us all a real scare, you know.

"He looks fine," barked the pilot on the left. He was not happy.

"Hey, kid. You know your poor mom is crying her eyes out over your little adventure. You should be ashamed of yourself. If you were my kid, I'd whoop you good."

I was suddenly sorry I had run up the steps.

"No need to be too tough on him, Jack." The co-pilot took up for me.

"Want some of my coffee, it's hot. It'll warm you up."

I took a small sip. It was foul. It was also warm. I chugged it down.

"Kid, get back downstairs," the pilot ordered. "We're taking off."

"Can I watch from here?" I asked from the steps.

The two pilots looked at each other. The pilot shrugged.

"Sure, kid," the co-pilot said. "You can watch us fly back if you do not move and you hold on. You must have had quite a time out here. You certainly earned your right to enjoy the ride back.

"You don't know how close we came to not finding you, kid. We were low on fuel and on our way back to the base. If you hadn't fired off that flare when you did, we would never have found you."

"Yeah, that's right," growled the pilot. "We weren't coming back out until daylight. Huntin' for a needle in a haystack in the middle of the night is crazy. You're damn lucky you fired off that flare when you did, kid. You would've frozen to death out here for sure. All right, hold on. Here we go."

The copter's engine revved up. In a moment, we were up and away. I was left wondering, how *did* they find me? I shot no such flare. Whatever they saw, it was not from me. I have asked my spirit guide that question many times since. And every time it has given me the same response—a Cheshire cat's smile.

17. WALKING TWO PATHS

All right, let's be mature adults about all this. We need to analyze what happened objectively without preconceived opinions. Two questions present themselves:

1. Were the spirit guides real or hallucinations?

My asthma was real. Both the girl who sat next to me in class and my own parents objectively observed the affect asthma had on me. After the injection of epinephrine, my heart-beat and breathing returned to normal. I was clear minded as I left the hospital, so I know I had survived a medical emergency. All this is objective reality—not the result of hallucination.

Further, even though connected by the umbilical cord to my oxygen-starved brain, my mind was clear enough above the lamps to experience frustration at not being able to control my "astral self" to position myself away from the surgical lamps so I could see better what was happening to my physical body on the treatment table. The desire to observe and witness what was happening to me on the surgical table below is evidence I was consciously on task and responding appropriately to it. This behavior is not consistent with someone whose mental faculties have been impaired.

Except for the epinephrine, I had taken no drugs or medicines that might affect my interaction with objective reality. The injection was not into a vein where blood flow would take the medicine into the brain and possibly distort perception. The young ER doc injected the drug straight into my heart.

The conclusion, therefore, must be that I did not create the appearance of the spirits or the portal. I experienced an objective reality that exists independent of our own and beings that exist independent of our own dimension of time and space.

2. Did the spirit guides influence the young ER doctor's treatment of me?

The male spirit guide told me the doctor had been given a puzzle that he might not be able to solve in time to save me. The "puzzle" the male spirit was obviously referring to was what was needed to restart my heart.

Why did my heart stop? Why was I not breathing? These are diagnostic questions I am sure the doctor must have been asking himself as he examined me. I don't recall the doctor asking my parents if I was suffering from an acute asthma attack. No one brought up the possibility of an asthma attack.

The doctor then simply began pushing my chest up and down and breathing into my mouth—what we now know as cardiopulmonary resuscitation (CPR). Without knowing I was suffering from an acute asthma attack, he would have no clue that CPR would not be enough.

The doctor's choice of treatment (simple CPR) indicates he had made the *wrong* diagnosis. CPR does not alleviate an acute asthma attack. Lack of oxygen into the lungs is *not* the problem. The problem is that asthma creates a biochemical blockage preventing the inhaled oxygen from being absorbed into the blood. I had all the oxygen I needed, but I couldn't use it until that biochemical blockage was removed.

Was that the "puzzle" the spirit guide had given the young doctor? It certainly seems like it must be. Without epinephrine, I die. No matter how many times or how well CPR is done—I die.

Unless the doctor came to the correct diagnosis, the remedy (epinephrine) would never occur to him. The doctor's test was to figure out I was suffering from an acute asthma attack with no physical evidence except my physical appearance. He had no medical history to give him any clue.

Until I told the female spirit guide I did not want to go with her, the doctor could not figure out what was wrong with me. But as soon as I tell her I want to stay, the doctor suddenly "knows" the answer. While it is possible these two events just happened coincidentally at the same time, it is much more likely the doctor's initial confusion and then his sudden understanding were both affected by the spirit guides.

What if the doctor's sudden "eureka moment" was not all his doing? What if even part of his sudden decision to inject epinephrine into my heart was not entirely his idea? My perception, at the time, was that the spirits were trying to get the young doctor to figure out that the drug I needed to live was epinephrine. I think they were planting thoughts into his mind, hoping he would catch on.

True, my interaction with the spirits was limited to when my brain was affected by a steady decrease in oxygenated blood. I was very close to death. My consciousness had separated from its physical host. In that condition of physical crisis, it makes sense that if spirits existed, I would encounter them.

The doctor was not so impaired, however. Assuming the spirits did plant the suggestion to the doctor to inject me with epinephrine, they exposed to me their ability to plant thoughts in the minds of ordinary humans at any time they choose.

As additional evidence of authenticity, consider my medical condition before and after my near-death experience. Before my NDE, I could not ride my bicycle or even walk much without becoming short of breath. Afterward, my health improved dramatically. Before long, I was riding my bicycle from our house on Pinchot Avenue to East Thomas Road and back without breaking a sweat.

Asthma would remain, but over time, the acute attacks came less and less frequently as my lungs gradually became stronger and healthier. By the time I reached adulthood, no one would ever guess that I had been so sickly a child. By age eight, I was playing organized baseball. By age 10, I was in Little League as a talented catcher.

But the changes were not merely physical. My encounter with the spirit guides changed me mentally and spiritually as well. Before my encounter, I accepted all the so-called Christian mysteries at face value—as any young child would.

Jesus as the only son of God?

No problem.

Mother Mary as a Virgin Mother?

Cool.

Walk on water? Water into wine? That Jesus never married or had a sexual relationship with anyone?

Got it.

Jesus rose from the dead and now stands upon the throne of God forever and ever, amen?

Amen, to that, too.

But, after meeting the spirit guides, nothing was simple anymore. I began to take my spirituality more seriously than any one of my peers. Surely, that is understandable, given what had happened to me.

As a result of my experiences, I began to question everything I was taught by the church.

No longer was I a blind believer. An inner "eye" had been opened.

Where other children my age were enjoying the simple pleasures of Christian fellowship and memorizing Bible passages, I was questioning the validity of many of the so-called "holy mysteries" upon which Christianity's alleged primacy as a religion is based.

I began to compare Christian teaching with what I had experienced with the spirit guides.

Too self-conscious to share my growing doubts with my parents or friends (they were all strong Christians), I asked myself questions. For example, one such question I spent a lot of time on was:

Was my encounter with the spirits consistent with my church experience?

I had to admit it was not. For example, some churches are full of golden relics, golden crosses, golden candelabra, golden this, golden that—everything golden. It seems some churches believe there must be some type of interdenominational competition to see which church possesses the most gold and silver. It is ironic how the institution preaching the values of spiritual simplicity and anti-materialism loves to flash its material earthly wealth. Who is trying to con who here? This hypocrisy is so obvious, even a child can see it.

None of the spirit guides I saw wore any clothing. All I saw were pale white orbs of light. They wore no gold, no silver to impress me. They did not wear immaculately woven robes like our clergy does. They carried no golden cross, no Star of David, no crescent moon.

Did any of the spirits proclaim they were affiliated with any religion?

No, they did not. When the spirit guides appeared out of their portal, none proclaimed their presence was in the name of any religious figure such as Christ or Jesus or even Buddha. They did not even announce their presence in the name of God. During my entire time with the spirit guides, I never heard them use the name "God" at all.

The female spirit guide I encountered during the car ride to the hospital knew I was in a Christian household. She did not mention it, but I could tell she knew me well. Just as I can tell what sport a person plays from the type of uniform they wear. A soccer uniform does not look at all the same as a baseball uniform.

Though she obviously knew my parents and I were Christians, the female spirit never once used Jesus's name to persuade me to come with her. In fact, in all their attempts to persuade me to go with them, the spirits never used my religious indoctrination against me whatsoever.

None of them told me that if I was a "good Christian" that I should go with them.

One would think that a 6-year-old would submit to Christian manipulation had the spirit guides wanted to do that—but they didn't try that tactic. Therefore, I am left with the conclusion that their interactions with us are regulated by certain ethical constraints. For example, they apparently are not allowed to intrude upon our inherent right to free will.

Could it be they respect their own code of ethics more than our earthly religions? What does that say about our earthly religions? It tells me the spirit guides value our individual, human spirituality above any earthly religion. It also tells me they respect the individual spirituality of others more than religions do.

The spirits obviously do not honor any religious ideology we mortals worship here on Earth. They appear to operate independently of all human religions. Our religious beliefs are of our own making—not theirs. Whatever their purpose may be in the larger scheme of things, the spirits show no connection with any of humanity's religions.

What should we make of this? For me, this was a life-altering revelation. It is said that "where there is smoke, there is fire nearby." Even as a youth, I knew that where hypocrisy lives, fraud hides nearby.

I began to walk two paths. The first was the path of Christian doctrine, where I tried to believe the religious indoctrination I was being taught through my church and my catechism. My Christian values were strong, reinforced by my churchgoing parents and Christian friends who would never have understood the ongoing struggle within my mind and soul.

The second path was a far different reality. On this path, I questioned how I could contemplate being a priest when I didn't even believe in the rituals I would be performing. How could I attack the church as hypocritical if I was willing to lie to myself?

My youth became a sea of turmoil both emotional and spiritual. My rational mind was at war with my emotional dependence upon the faith I had been raised in. Some days, I would be a "good Christian" striving hard to reconcile my doubts with my faith. Other days, I was the rebel questioning everything, fighting everyone.

Would one side ever overcome the other and grant me peace? If that happened, which side would it be? As a teen and then a young adult, these were troubling times for me. My indoctrinated mind convinced me that at stake was nothing less than my immortal soul. I began to believe there was no resolution, no peace, no certainty for me. I feared I would never be able to cope with the internal conflict afflicting my every waking moment.

Fortunately, I had begun meditation when I was very young. Out of meditation, I learned to stop fighting with myself about the conflict. I put the issue in the hands of those who were more qualified than I and let them handle it—my spirit guides. This became what I would come to call my "Third Option." I would not worry anymore about what to do about religious hypocrisy and how it affected my personal life choices. I would let my spirit guides decide for me what I needed to do about it. I trusted them implicitly. If it matters, I still do.

After earning my Master of Arts degree in International Relations at Kent State University's Graduate School, I dedicated a full month of meditation to whether I would attend seminary or law school. I wanted spiritual guidance to make my choice. What would my spirit guides want me to do?

But they gave me no clear guidance. They did not tell me what to do. I faithfully did my daily meditations, but no guidance, no roadmap, no direction came. The month of meditation neared its end, and I was no closer to an answer then than I had been at the beginning. Frustration was setting in.

Then, on a trip to a bookstore, Mohandas K. Gandhi's autobiographical *The Story of My Experiments with Truth* caught my eye. On a whim, I bought it. Once I started reading, I could not put the book down.

Gandhi's book was a personal revelation. He was brutally candid about his own human weaknesses. He pulled no punches about his shortcomings. By so doing, however, Gandhi displayed a strength of character and spirit that made him a lion among mortals. He became my personal hero. He still is.

This frail little man forged the revolution of India's independence, bringing mighty Britain to its knees. All he did was to teach how to be true to one's self by being humble and having one's thoughts be consistent with one's ideals. That is all it took.

His secret was as simple as it was effective. One need not use a gun or other weapon to force change. One need only to pursue loving and nonviolent resistance, and, eventually, the oppressor's behavior will change. He believed that by sharing the spirit of love within, he could eventually change the heart of even the most brutal dictator. Where the soul goes, the mind will surely follow.

Gandhi's life story teaches that by strength of character and spirit one can change the course of history. When one's spirit is cleansed of hate and fear, there is no longer any need for violence to affect change.

Gandhi's autobiographical story changed my life. I realized I did not need to be a priest to be true to my ideals. Gandhi was no priest. He was a lawyer, educated in the law in England before returning to India and eventually to immigrating to South Africa, where he would first experience racist discrimination and vow never to accept such evil.

Gandhi's life story gave me the inspiration to be true to my own soul. Instead of enrolling at a seminary and entering the priesthood, I applied to law school at The George Washington University's National Law Center in Washington, D.C.

18. THE DREAM

It is late August 1979.

Jimmy Carter is president of the United States. That January, Brenda Ann Spencer murders two teachers and wounds eight students at a school in San Diego, California. In February, Muslim extremists kidnap Adolph Dubs, the American ambassador in Kabul, Afghanistan. Dubs is killed in a shootout between the terrorists and authorities soon thereafter. That March, the space shuttle Columbia *is prepared for its first launch at the Kennedy Space Center in Florida.*

That April, the Albert Einstein Memorial is unveiled at the National Academy of Sciences in Washington. That July, President Carter approves secret aid to the mujahideen of Afghanistan to fight the Soviet Union's occupation there. In early August of 1979, Michael Jackson releases his album, Off the Wall.

The year 1979 is also a year of profound challenge for me personally. This is the year I race headlong, blindly into the single most important crossroads of my life.

Would I succeed?

Or would I fail?

Classes as The George Washington University's National Law Center in Washington, D.C., were to begin the day after Labor Day of that year. I had, unwisely, waited until the week before classes started to get to D.C. to find a place to live.

There are times when wisdom must take a back seat to the heart. The reason I was late getting to Washington was a beautiful young nursing student I had fallen in love with during my last year at Kent State. I knew I risked not finding anywhere to live if I waited too long, but I just couldn't leave her.

We were both very happy with each other. Neither of us wanted me to leave for law school. Our time together was too idyllic, so we just kept putting off the inevitable. Confident I would find something at the last minute, I stayed with her until the week before classes started.

When I finally arrived in Washington, a rude awakening awaited me. There were no apartments available. I looked throughout Arlington, Alexandria, Dupont Circle, and even Maryland for a place to live.

And failed.

There was nothing available. I had waited too long. By the time I arrived in D.C., most of the student apartments around the city were already taken. The only apartments left, were much too expensive for me.

I began attending law classes in the day and living out of my suitcase at night. I was basically homeless. My situation had "DISASTER" plastered all over it. Desperate, I trudged my suitcase and law book–filled backpack downtown.

Located in Washington, D.C., on 11th Street Northwest, the International Youth Hostel is what its name implies—a place where students traveling from different countries can stay that is safe and inexpensive. I wasn't going to get any studying done until I found somewhere permanent to live, and those were the two qualities I needed immediately. Only three blocks from Washington's notorious red-light district, the hostel was a safe haven amid the chaos of Washington's urban life. It gave me what I needed—a safe, clean place to wash up and sleep. It was also cheap at only five bucks a night.

But, there was zero privacy. I shared one room with three other guests, triplets from Amsterdam, The Netherlands. Each of us slept in a military-style bunkbed and shared a single bathroom, which was down the hall.

The street noise outside my window was enough to keep me awake by itself. Downtown D.C. was sirens, screaming, and shouting 24/7. But what was worse were my three, happy-go-lucky roommates.

I never got to know their names. They spoke no English. They spent all night, every night, jumping up and down in the room to techno-pop at max volume out of their three boom boxes—all the while waving brightly colored glow sticks and light wands around in the dark. By the second night, I thought I had gone mad.

Every night was an insane rave party. Every day, I went to law classes less awake than the day before. My brain was fried. The noise was driving me crazy. If something did not happen, and soon, I would have to withdraw from law school and pay back a pointless loan.

Worse, I would be an idiot for letting my love life ruin my future. Who lets their love life blind them to the practical necessities of real life? I would come back home to my parents worse than a failure. I would return as a fool.

It was a Tuesday night. The tuition-refund deadline was that coming Friday, but I was not going to wait that long to leave school. Sirens and screams assaulted me from outside. The techno-pop triplets blasted me with their blaring noise from inside.

I'd had enough. The next day, Wednesday, would be my deadline day. If I didn't find a place to live by 5 p.m. the next day, I would withdraw from classes on Thursday. Friday would be my go-home day.

And on Saturday?

I did not even want to think about what would happen after I went home. My mother would not say the words, of course. Rather she would think them all the same: "You failed." Then she and my grandma would try to convince me to settle down around town. But without a meaningful job there would be no meaningful career. I would not be relevant. Life would pass me by without so much as a nibble from me. I would lose my mind.

All that as a consequence of a summer I could have spent finding a place to live, but I wasted instead on a college love. Well, here I am, and look where it got me, I thought to myself at the time. It would take a miracle for me to be able to stay.

Sitting on my bunk, I watched the techno triplets do their thing. It was past midnight. I should have been asleep already with class the next morning, but I didn't care anymore. The lights were turned off, so they could wave their orange and green Day-Glo light-sticks in the dark. I had to admit it was beautiful to watch their intricate orange and green patterns in the air as they danced.

"Do you still believe in miracles?"

It was the voice. It was back. I almost jumped out of my bunk.

I had not expected to hear that voice. With my law classes during the day and being driven to insanity at the hostel each night, I had not even thought of the voice since before I left Kent State, where I did my undergraduate studies. Immediately, I laid down in my bunk and drew the pillow over my head to drown out the noise. I begged my spirit guides for a miracle. I told them everything.

I told them I screwed up by not getting an apartment in time. Unless I found one the very next day, I would have to withdraw from law school. If they wanted me to be a lawyer, they needed to help me and help me now. It was up to them. I was helpless.

Would they answer? Would the voice tell me what to do?

I waited. Pillow pressed firmly over my head to drown out the triplets, I waited for a sign, a word, anything.

They gave me silence, instead.

The noise from outside and inside the room faded away.

I fell into a very deep sleep.

That night, I had a dream—what is called a waking-dream. A waking-dream is the kind one has while asleep, yet the person is totally aware of the events that occur during the dream. These dreams are more real than any virtual reality game I have ever played. They are as real as life itself. In fact, I believe they are actual life experiences from the past and the future. I have had many such dreams in my life, and I remember them all.

****

I am standing just outside Stockton Hall, the law school's main building, near the corner of 20th Avenue and "G" Street Northwest. The sun is going down, so I know it is around 4:30 in the afternoon.

The sidewalks are full of GWU students going home after classes, and I am waiting for the light to change.

In the dream, I start walking north on "G" Street toward the Foggy Bottom metro stop. To my left is an older red-brick school house with a chain-link fence protecting the asphalt-covered play area inside. As soon as I reach the school, some external force twists my head to the right. Across the street is a one-story brick building. It is painted white. Inside is a small delicatessen.

Next to the deli is a small park bench by the street. Lying on the bench is what appears to be a pile of trash. As I stare, the grossest-looking homeless person I have ever seen sits up and waves at me. Matted gray hair falls past his waist. His stench is so bad, I can smell him from across the street. This poor fellow is so gross, he makes Aqualung look like a prince.

I recoil from him. He responds by urgently waving me over to him. He wants to meet me.

An unseen hand touched my shoulder. A force external to mine directed my sight away from the homeless man to a small park, immediately to the deli's right. On the other side of the park stood a two-story brownstone that looked like a private residence. Outside the brownstone's wooden front door was an aluminum screen-door. On the bottom half of this aluminum screen-dorm a large "T" was engraved into the metal frame along with the number: 2163.

"There," the voice commands. "Go there."

****

I woke up.

It was dawn, already. My daily regimen kicked in, and I robotically got ready for my first class. By the time I made it to class that morning, the dream was totally forgotten. By the end of class that afternoon, I was so consumed with legal issues from my torts and contracts classes, my mind was mush.

I walked out of Stockton Hall mindlessly, heading nowhere in particular. By the time my mind-fog lifted, and I looked around to get my bearings, I found myself standing on the corner of 20th Avenue and "G" Street Northwest waiting for the light to change.

What happened next was a precise repeat of the dream—only in real life. It all happened just as the dream foretold.

Walking down "G" Street Northwest, I saw the red-brick public school house with the chain-link fence. On the other side of the street is the small white-brick building with the deli inside. Just in front of the deli is a park bench. Relaxing on the bench was Agualung. He waves at everyone. He sees me. Now he waves at me, too. I wave back.

The building immediately to Aqualung's right was a two-story brownstone. An aluminum screen-door with a "T" on it adorned its front entrance. Its address:

2163 "G" St. NW.

Heart pounding, mind reeling, I crossed the street from the school and walked up the steps to the door with the "T" on it. My hand trembled as it rose to knock on the door. I took a deep breath.

Then I took another.

My dream to be a lawyer hung in the balance. All the hard studying, working, and planning I had diligently dedicated myself to while at Kent State now depended upon something completely out of my control: whoever answered the door and how they reacted to me.

My hand steadied long enough to form into a fist.

Knock…

Knock…

Two young women answered the door. They were both GWU graduate students: one blonde, the other brunette. Both were stunningly beautiful. The house from my dream just happened to be the only GWU graduate group residence on the block.

"Do you have an opening here?" I blurted, desperately trying to be charming but not so much as to appear phony. My heart was in my mouth. It was difficult to speak.

"I'm a new graduate student. I have been looking for over a week to find a place to stay. If I can't find something soon, I'll have to leave school. I'm desperate."

"You must be a law student," the brunette sniffed, frowning. "I don't like law students. Most of you are just a bunch of assholes."

I had no idea how she could tell I was a law student. Could it have been the three-piece pin-stripe suit I was wearing? The law school required all its students wear a business suit to class. She knew law students by how they dressed. My heart sank.

"You have to excuse her," the young blonde woman said as she stepped between us. She offered an encouraging smile. The brunette faded away behind her.

"She's just had a bad day. What's your name, sweetie? Come on in. What she says about most of the law students we run into around here is true, but I can tell you're not like the rest. We just happen to have a room opening up in two weeks. If you can handle living in a closet in the basement until then, the room is yours."

19. THE QUANTUM SEA

What was behind my dream? Who was behind that dream? These are the questions that should concern us. It certainly did not come from my own consciousness. Nor did it originate from within my super-consciousness or unconscious mind. It could not have.

First: I had never been to that group house before in this life.

Second: I had never walked down that street before.

Third: Before the dream, I had never, ever seen Aqualung.

Was it a Christian dream? The dream did not come with any of the bells and whistles one expects of a "Christian miracle." Jesus did not appear—or Satan. Not once, in all my dreams over the years, have my spirit guides given me any indication they were affiliated with Jesus, or the devil or anyone connected to either of them.

The dream was not religious in any sense of the word. The dream's purpose was pragmatic rather than mystical. Its purpose was not vague, it was specific—to keep me in law school.

This dream was planted in my subconscious, at the waking-dream level, so that I would find a place to stay during law school. My spirit guides wanted me to become a lawyer. There can be no doubt about that.

Had the spirit guides come to me as angels in the Christian tradition at that time in my life, I might have seen it as a Christian miracle and quit law school and gone to seminary instead. In retrospect, the spirit guides went out of their way to avoid any religious iconography that might "flip the switch" to awaken the still-smoldering embers of my religious indoctrination.

The Washington, D.C., dream is compelling evidence the two events I experienced as a child in Phoenix were not hallucinations from oxygen starvation or drug-induced delusion. My near-death experience in the hospital and the voice in the desert were, therefore, real. The events happened as I perceived them. They were not creations of my own mind or my imagination.

Therefore, there is but one explanation. Both the Washington, D.C., dream and the Phoenix experiences were not the creation of my own mind or imagination. The events I experienced were independent of my conscious perception. This leads us to but one conclusion: a being or beings unknown to us were controlling these experiences.

Who are they? What are they?

I admit I do not know. Think me crazy if you want, but these experiences happened to me as I have related them to you, the reader. You may choose to ignore the implications, but I have been forced to live with them every day of my life.

Likewise, PhiAlpha compels certain implications that cannot be avoided. Evolution has no conscious intention of its own. The consciousness of each of us is connected to memories from lifetimes we may not remember consciously but which exist nevertheless. These experiences are available to all of us—if we open our minds to them. The knowledge of who we really are as beings is kept in our memories. They are the seeds of our personalities, fertilized in the cosmos eons ago and nurtured over innumerable lifetimes.

Accessing past-life memories is not a gift. Each of us has the potential to experience them. There is no hidden trick. What is necessary is a conscientious intention to open one's mind to these memories fearlessly. One cannot meditate in fear of demons and evil and be successful. One must realize demons and evil are human creations passed on from generation to generation for the sole purpose of scaring people from pursuing spiritual growth independent of the church.

Let's be real. Without fear of demons and hell, there is no need for the church. Do you not believe the church knows that? Move on with your spiritual growth independent of the fears implanted within your mind by the church and see how quickly you grow within.

Key to this growth is meditation. Meditation is the key to opening one's consciousness to the memories hidden underneath and the knowledge contained within. Meditation also opens our minds to universal knowledge and wisdom—and a sense of ever-growing unity with all life.

The moment we open our consciousness up to these memories, our lives change. We begin the download of memories of who and what we truly are.

By the way, remember the homeless man from the dream who stank worse that Aqualung?

He was a graduate student from one of the university's science departments. Allegedly, he had an IQ comparable to Einstein, had almost a hundred patents to his name, and was independently wealthy.

He was fascinating to talk to, as long as you could ignore his body odor—which was stomach-churningly horrific. The trick was to stay upwind. Over the next three years, we spent hours together on that park bench discussing everything from his view of world events to why he hated his mother. According to some on campus, he was one of the brightest minds of the 20th century.

However, he suffered from schizophrenia. He loathed people. To avoid them, he turned himself into a fictional character to literally keep them at arm's length. He became a real-life Aqualung—and watched the world go around from his perch atop his own "Fool's Hill"—the name he gave for the stoop he would sit on watching all the students and federal employees walk by every day.

Where did my dream come from? Did it come from God? Did it come from my guides who have been watching over me all my life? Or, did it come from neural download of memories from a past life as a result of PhiAlpha? Does this dream, in and of itself, provide the proof of PhiAlpha we seek?

These are mind-blowing questions, to be sure. One thing is for certain, without a serious scientific effort to find out once and for all put the PhiAlpha theory to the test, we shall never know.

20. THE NEXT STEP

PhiAlpha's existence challenges us to cast aside what we thought we knew about life and death. PhiAlpha challenges us to reconsider what we experience every day as free will or predestination. Is creation static and unchanging? Or is creation an evolutionary process—an endless spiral of infinite realization?

As an example, let us take the issue of whether God created the heavens and the earth and humans in six days as it is written in the Bible's Book of Genesis. We are taught to believe that the universe must, therefore, be complete. Further, because "man" is part of the universe, we are taught that man is complete—but by sin, flawed.

With this doctrine, we are the equivalent of spiritually indentured servants who are given only one path to freedom (heaven)—through the church. To earn a place in an afterlife of heaven, we must first kiss the ring of religious doctrine.

However, PhiAlpha tells us a much different story. PhiAlpha tells us that creation is an ongoing process during which nothing is fixed and all can become enlightened. PhiAlpha tells us the Big Bang was just the first note of an eternal symphony of creation and self-realization.

The universe is in the process of creation every moment of every day. The force or forces that created evolution are of life. Inherent within life is the freedom to adapt and change to survive and flourish—both spiritually and physically.

The galaxies did not just appear and then stop in place at the Big Bang. No, the galaxies, their stars, planets, and the beings living on them have been in motion ever since. creation is as infinite in time and space as the universe itself. Thanks to former Bell Lab's Robert Wilson and Arno Penzias, we know the echo of the Big Bang can be heard, even today, as cosmic background radiation. creation has not ended. In truth, the human chapter of creation has just begun.

Consider, for example, the idea that since we have not yet been in knowing contact with an alien civilization, that we must be alone in the universe. This is akin to saying the world is flat.

It took primordial life on our Earth billions of years to develop—and billions of years longer to evolve into the sentient beings we have become. Are we that much different from the rest of the cosmos? If it took life that long to develop here, what says alien races evolved faster? Where is the evidence it took less time anywhere else?

There isn't any such evidence. We have yet to discover life beyond this planet, let alone advanced civilizations. Therefore, for now at least, we here on Earth are the best evidence of how universal creation and life interact.

PhiAlpha takes us one step further. PhiAlpha says our path to a new life is no longer dependent upon religious doctrine at all. PhiAlpha shows that we have infinite lifetimes—not because we are good or bad, saved or unsaved—but because our spirits are eternal. If you can get to the next life without believing in a god—what is the point of the clergy or the church? What other product can they sell us that we would buy?

What is the next step?

If PhiAlpha makes present Christian doctrines irrelevant, does this mean the church itself is now irrelevant? Like any other human organization, the church suffers its share of internal political thuggery and mischief. The organization of the church will never be perfect because it will always be subservient to imperfect mortals.

PhiAlpha teaches that we are in control of our own individual spiritual destiny. We have always had the power to do so. The church simply hid the truth from us. PhiAlpha tells us we must take over our own spiritual destiny. But first, we must accept the mantle of authority to judge ourselves and each other as part of our communal oneness within the spirit of creation of which PhiAlpha is a part. This is the lesson of self-accountability and redemption through mutual forgiveness and good works the world needs today.

PhiAlpha teaches that we must stand on our own two feet spiritually. We must be willing to accept responsibility for ourselves and our communal (local, state, and national) behavior in the world. We must become our brother's keeper, inspiring each other to seek greater meaning and personal enlightenment while also demanding individual accountability at all levels of society.

PhiAlpha is the beginning of our spiritual renaissance, but it is not the end—far from it. Humanity has yet to figure out the whole "spiritual deal"—just as it has yet to figure out many of the countless mysteries of the universe.

We do not know—I do not know—exactly *who* it was that spoke to me in the desert near Camelback Mountain or what gave me the dream that saved my law career. Does it not disturb you, the reader, not to know? It bothers me, and I have lived with these spirits all my life.

Please do not tell me the sophomoric answer that they are "angels of God." With all due respect, that answer is no longer acceptable. There is no thought to such a conclusion. There is no intellectual analysis. We are no longer in spiritual first grade. More is required of us now.

It is not an attack upon faith to insist we know upon what rock our faith stands. If the spirits I encountered were part of the "Army of the Archangel Gabriel," I am confident they would have told me. But they didn't. So, I don't think the old concept of angels applies. It is fake news.

We are dealing with entities that exist in real space and in real time. They have been proven to affect physical and biological processes without being corporeal themselves. They are not mythical creatures from ancient times but real-life beings that interact with us every day.

Do we know their origin? Do we know their intentions? Do we know how they are able to influence us—often, I believe, without our being aware of it?

The answer to each of these questions is "no." But, at least, these mysteries are real ones and not fabricated by a religious sect for personal wealth, hubris, or power. We have many questions, and there are still few answers that we can cling to with confidence.

But the answers we can cling to are of science and fact. Each new path begins with one small step after another. Phialpha is but one of many steps we will continue to take together as we walk the path of life one day, one discovery, at a time.

What is the next step?

The existence of PhiAlpha demonstrates that our DNA is constructed to allow our consciousness to become more complex with each generation. Consciousness is a commingling of classical and quantum mechanical processes affecting our intellect and emotion, our physiology and spirit, and our rational and subjective perceptions. All this information is integrated within our individuality through PhiAlpha.

Does PhiAlpha recognize the existence of a soul? Of course, it does. But let's be careful as to how we define "soul." The soul PhiAlpha recognizes is that part of the human mind that is emitted within the biophoton capsule as the body dies. This is the part of us that survives death and passes on our individuality to a new generation of physical bodies.

PhiAlpha also establishes that each individual soul was created to become increasingly enlightened with each rebirth. The soul is an integral part of our consciousness. The raising of one's consciousness enlightens both the intellect and the spirit as well. The biophoton capsule of memories carried over from our past life becomes the seed out of which our individuality will be resurrected in the next life.

Enlightenment is impossible, however, until the mind is freed from the self-imposed chains of limitation and the external chains of doctrinal intolerance. While some explore enlightenment on their own, most remain spiritually bound by religious doctrines that are 1,700 years old.

Concepts that were created to control the interaction of simple merchants, hunters, fishermen, warriors, and farmers have little meaning to us now. And, do not forget the way religious doctrines were used to control women. Women were regarded as little more than property—and often much worse.

The power of mind control cannot be doubted. Two thousand years have gone by, and these same institutions still control the behavior of millions in the world today. That must change. Enlightenment of the intellect and the spirit must be more than a hobby for the bored intellectual elite of the world. Unless freedom from mind control is made available to the masses, the gains made will be slow and only with grudging resistance.

What is the next step?

The next step is liberation of the human soul. PhiAlpha is the beginning of a new reality, a new vision of humanity's purpose in the universe. With PhiAlpha our innate spirituality can rise out of our subconscious, free of the artificial constraints of Christianity's myths, to finally stand as inheritors of our divine purpose.

REFERENCES

1. Craffert, P.F., "Do out-of-body and near-death experiences point towards the reality of non-local consciousness? A critical evaluation," *Journal for Transdisciplinary Research in South Africa*, 2015, Retrieved at dspace.nwc.ac.za.

2. Borjifin, J., Lee, U., et al, "Surge of neurophysiological coherence and connectivity in the dying brain," Proceedings of the National Academy of Science (PNAS), 2013, Retrieved at cbi.nlm.nih.gov.

3. Chawla, L. S., "Surges of electroencephalogram activity at the time of death: A case series," 2011, *Toward a Science of Consciousness: Brain, Mind, Reality*.

4. Lake, James H., "The near-death experience: A testable neural model," 2016, *Psychology of Theory, Research and Practice, 4(1), 115-134*.

5. "All about entropy, the laws of thermodynamics and order from disorder," 2001, *Archives of Science*, Retrieved at entropylaw.com.

6. id.

7. id.

8. "What is thermodynamics," NASA Glenn Research Center, (n.d.), Retrieved at grc.nasa.gov.

9. Zhang, B. and Cai, Q., et al, "Information Conservation is fundamental: recovering lost information in Hawking radiation," Retrieved at arVix, 2013.

10. Horodecki, M., Horodecki, R., et al, "No-deleting and no-cloning principles as consequences of conservation of quantum information," arXiv, 2003.

11. id.

12. id.

13. Roncaglia, M., "On the Conservation of Information in Quantum Physics," arXiv, 2017.

14. Samal, J. R., Pati, A. K., and Kumar, A., "Experimental Test of the Quantum No-Hiding Theorem," *Physical Review Letters*, 106, 2011.

15. Shi, S., Kumar, P., and Lee, K.F., "Generation of photonic entanglement in green fluorescent proteins," *Nature Communication*, Dec. 5, 2017.

16. Tentrup, T., et al, "Transmitting more than 10 bit with a single photon," arXiv, 2017.

17. "Information in a Photon," (no author), DARPA (Defense Advanced Research Projects Agency, (no date).

18. id.

19. Noble, P.A., Pozhitkov, A.E., et al, "Thanatotranscriptome: genes actively expressed after organismal death," 2016, *Open Biology*, Retrieved at biorxiv.org.

20. id.

21. Jacobs, G., "An evolutionary view of individuality," 2012, *World Academy of Art and Science,* Retrieved at worldacademy.org.

22. Smith, John Maynard and Szathmary, Eors, *The Major Transitions in Evolution*, 1995, Oxford University Press, Oxford, England.

23. West, S.A. and Fisher, R.M., et al, "Major evolutionary transitions in individuality," 2015, PNAS, 112(33), 10112-10119, Retrieved at ncbi.nlm.nih.gov.

24. Michod, R.E., "Evolution of individuality," 1998, *Journal of Evolutionary Biology*, 225-227; See also Michod's many other articles regarding the evolution of individuality including

"Evolution of individuality during the transition from unicellular to multicellular life," 2007, PNAS, 8613-8618.

25. id.

26. McLean, C.Y. and Reno, P.L., et al, "Human specific loss of regulatory DNA and the evolution of human-specific traits," 2011, *Nature*, Retrieved at nature.com

27. Marino, L., Dunbar, R., Lahn, B., and Preuss, T., all cited in: Bradbury, J., "Molecular insights into human brain evolution," 2005, *PLOS Biology*, Retrieved at ncbi.nlm.nih.gov.

28. id.

29. id.

30. id.

31. Sherwood, C., et al, "A natural history of the human mind: tracing evolution changes in the brain and cognition," 2008, *Journal of Anatomy*, Retrieved at ncbi.nlm.nih.gov.

32. id.

33. White Matter. (n.d.) In Wikipedia.

34. Delude, C., "Researchers show that memories reside in specific brain cells," 2012, *MIT News*, Retrieved at news.mit.edu.

35. Stufflebeam, R., "Neurons, synapses, action potentials, and neurotransmission," 2008, *Consortium Cognitive Science Instruction*, Retrieved at mind.ilstu.edu.

36. Harris, K. and Sejnowski, T., et al, "Nanoconnectomic upperbound on the variability of synaptic plasticity," 2015, *eLife*, Retrieved at elifesciences.org.

37. id.

38. Neurons. (n.d.) In Wikipedia.

39. Chudler, Eric H., "Neuroscience for kids," 2016, University of Washington Press, Retrieved at faculty.washington.edu.

40. Neurons. (n.d.) In Wikipedia.

41. The brain core. (n.d.), The Brainwaves Center, Retrieved at brainwaves.com.

42. No author, "Amygdala and Hippocampus," In Wikipedia, date.

43. No author, "Stria Terminalis," In Wikipedia, no date.

44. No author, "Congulate Gyrus," In Wikipedia, no date.

45. No author, "Corpus Callosum," In Wikipedia, no date.

46. No author, "Fornix," In Wikipedia, no date.

47. No author, "Hypothalamus," In Wikipedia, no date.

48. No author, "Indusium Griseum," In Wikipedia, no date.

49. No author, "Septal Nuclei," In Wikipedia, no date.

50. No author, "Thalamus," In Wikipedia, no date.

51. No author, "Spinalthalamic Tract," In Wikipedia, no date.

52. Huang, Y., et al, "Measurements and models of electric fields in the in vivo human brain during transcranial electric stimulation," Retrieved at eLife 2017:6:e18834. See also: Irani, David, "Cerebralspinal Fluid in Clinical Practice," Chap.10, 69-89, Saunders, 2009. Agarwal, R.P., Henkin, R.I., "Zinc and copper in human cerebralspinal fluid," *Biological Trace Element Research*, 1982 June;4(2-3); 117-124; and Lipkova, J. and Cechak, J., "Human electromagnetic emission in the ELF band," Radar Dept., Faculty of Military Technology, University of Defense, Czech Republic, in *Measurement Science Review*, Volume 5, Section 2, 2005.

53. Understanding the stress response. 2011, *Harvard Health Publications*, Harvard University, Retrieved at health.harvard.edu.and "The Brain from Top to Bottom" (n.d.), McGill University, Canada, Retrieved at thebrain.mcgill.ca.

54. id.

55. Sargis, Robert, MD, PhD, "An overview of the hypothalamus: the endocrine system's link to the nervous system," 2015, endocrineweb, Retrieved at endocrineweb.com.

56. Harvard, supra.

57. Cortisol Initiates. (n.d.), In Wikipedia.

58. id.

59. id.

60. id.

61. Mashour, G., Borjigin, J. and Lee, U., "Surge of neurophysiological coherence and connectivity in the dying brain," 2013, PNAS, Retrieved at pnas.org. Also See Noble, PE., et al, "Tracing the dynamics of gene transcripts after organismal death,"

Open Biology, 2017, Jan. 7 (1), 160267, Retrieved at rsob.royalsocietypublishing.org.

62. Mastin, L., "The human memory," 2010, Retrieved at humanmemory.net.

63. id.

64. id.

65. id.

66. id.

67. Becker, J.S. and Zoriy, M.V., et al, "Imaging of copper, zinc, and other elements in thin section of human brain samples (hippocampus) by laser ablation inductivity coupled plasma mass spectrometry," *Analytical Chemistry*, 2005, May 15; 77(10): 3208-16, Retrieved at juser.fz-juelich.de.

68. Angelova, M. and Asenova, S., et al, "Copper in the human organism," *Trakia Journal of Sciences*, Volume 9, No. 1, 88-98, 2011.

69. Sofronescu, A., PhD, "Cerebral spinal fluid analysis," 2015, Medscape, Retrieved at emedicine.medscape.com.

70. Cerebralspinal fluid. (n.d.), In Wikipedia; See also, Klarica, M. and Oreskovic, D., "A new look at cerebralspinal fluid movement," 2014, Fluids Barriers CNS.

71. Bauman, S. and Wozny, D., et al, "The electrical conductivity of human cerebralspinal fluid at body temperature," 1997, *IEEE Transactions on Biomedical Engineering*, 220-223, Retrieved at ieeexplore.ieee.org.

72. "Ion Channels," (n.d.), Biology Reference, Retrieved at biologyreference.com; See also Ion Channel. (n.d.) In Wikipedia.

73. Dolcos, F., Lordan, A., and Dolcos, S., "Neural correlates of emotion-cognition interactions: A review of evidence from brain-imaging investigations," 2011, *Journal Cognition Psychology*, 669-694.

74. Kennsinger, E. and Lorkin, S., "Effect of negative emotional content on working memory and long-term memory," *Emotion, 2003 Dec;3(4), 378-93*.

75. Amaral, D. and Scharfman, H.E., et al, "The dentate gyrus: Fundamental neuroanatomical organization (dentate gyrus for dummies)," 2007, Progressive Brain Research, Retrieved at

ncbi.nlm.nih.gov.

76. "The puzzling role of biophotons in the brain," 2010, *MIT Technology Review*, Retrieved at technologyreview.com.

77. Rahnama, M. and Tuszynski, J., et al, "Emission of mitochondrial biophotons and their effect on electrical activity of membrane via microtubules," *J. Intergr. Neuroscience*, Volume 10, No. 1 65-88, 2011, Retrieved at arxiv.org.

78. id.

79. id.

80. Caswell, J. and Dotta, B.T., et al, "Cerebral biophoton emission as a potential factor in non-local human-machine interaction," *NeuroQuantology*, March, 2014, pp 1-11, Retrieved at neuroquantology.com.

81. Costa, J., Rouleau, N. and Persinger, M., "Differential spontaneous photon emissions from cerebral hemispheres of fixed human brains: Asymmetric coupling to geomagnetic activity and potentials for examining post-mortem intrinsic photon information," 2016, *Neuroscience and Medicine*, 49-59, Retrieved at scirp.org.

82. Rouleau, N. and Murugan, N.J., et al, "When is the brain dead? Living-like electrophysiological responses and photon emissions

from applications of neurotransmitters in fixed post-mortem human brains," 2016, *PLOS One*, Retrieved at journals.plos.org.

83. Wang, Z. and Wang, N, et al, "Human high-intelligence is involved in red-shift of biophotonic activities in the brain," 2016, PNAS, Retrieved at pnas.org.

84. Furusawa, A. and Sorenson, J.L., et al, "Unconditional quantum teleportation," *Science,* Oct., 1998, Volume 282, pp. 706-709, Retrieved at sciencemag.org.

85. Sun, Qi-chao and Mao, Ya-Li, et al, "Quantum teleportation with independent sources and prior entanglement distribution over a network," 2016, *Nature Photonics*, 671-675.

86. Valivarthi, R. and Puigibert, M.G., et al, "Quantum teleportation across a metropolitan fibre network," 2016, *Nature Photonics*, 676-680.

87. Einstein, A., Podolsky, B. and Rosen, N., "Can quantum-mechanical description of physical reality be considered complete," 1935, *Physics Review*, 777, Retrieved at journals.aps.org.

88. id.

89. id.

90. Photon. (n.d.), In Wikipedia.

91. Feynman, R., "Relation of wave and particle viewpoints," 1965, *Feynman Lectures, Vol. III, Quantum Mechanics, Chapter 2*, Caltech.

92. Balaguru, S. and Uppal, R., et al, "Investigation of the spinal cord as a natural receptor antenna for incident em waves and possible impact on the central nervous system," 2012, *Electromagnetic Biology and Medicine*, Retrieved at ncbi.nlm.nih.gov; See Also "Spine antenna pointing mechanism (APM)" (n.d.), MMA Design, Retrieved at mmadesignlic.com., MMA Design manufactures and sells a patent-pending antenna for space satellite-Earth communication. It mimics the human spine, using a stack of ten vertebrae-like metallic, pivoting elements, one on top of the other, with metallic coils serving as discs. The spine has a hole in the center running longitudinally through the coils (just like a human spine has a hole for the spinal cord). Instead of a spinal cord, a flexible RF coaxial cable or wave guide (analogous to the spinal cord) is routed through the center of the device. The purpose of this spine-shaped

antenna is to transmit and receive signals between spacecraft and ground stations during space missions.

93. id.

94. "Pregnancy: Week by week. (n.d.), Mayo Clinic, Retrieved at mayoclinic.org.

95. Median section embryonic brain, (n.d.), In Wikimedia.

96. Eisaman, MD and Fan, J., et al, "Single-photon sources and detectors," *Review of Scientific Instruments*, Vol. 82, 071101, 2011. Their article provides an excellent overview of the many

different types of single-photon detectors and their application.

97. Marsili, F., et al, "Single-photon detectors based on ultra-narrow superconducting nanowires," *Nano Letters*, May, Vol. 11, 2048-53, 2011.

98. Baek, B, et al, "Superconducting a-W$_x$Si$_{1-X}$ nanowire single-photon detector with saturated internal quantum efficiency from visible to 1850 nm," *Applied Physics Letters*, Vol. 98,

251105, 2011.

99. Chunnilall, Christopher, J. and Pietro, I., et al, "Metrology of single-photon sources and detectors: a review," *Optical Engineering*, Vol. 53, 081910, 2014.

100. Sagan, Carl, *The Demon-Haunted World: Science as a Candle in the Dark*, Headline, London, 1997.

ILLUSTRATIONS

Illustration 1..Page 60

    Courtesy Wikipedia.com "Anatomy-Nervous System"

Illustration 2..Page 60

    Creative Commons Share-Alike 3.0 License courtesy Caio Arroyo 2010

Illustration 3..Page 64

    Creative Commons Share-Alike 3.0 License courtesy Oxford University Medical School and dansmedschoolnotes.com

Illustration 4..Page 70

    Creative Commons Share-Alike 3.0 License courtesy Tangent LLC 2018

Illustration 5..Page 75

    Courtesy Freefoto by Ian Britton (no date)

Illustration 6..Page 80

    Creative Commons Share-Alike 3.0 License courtesy regalrealness.com 2018

Illustration 7..Page 81

    Creative Commons Share-Alike 3.0 License courtesy Wikipedia.com

Illustration 8..Page 87

    Creative Commons Share-Alike 3.0 License courtesy Fatima Al Sayed 2013

Illustration 9..Page 98

    Pinterest 2018

Illustration 10..Page 99

　Creative Commons Share-Alike 3.0 License courtesy sharonapbiotaxonomy via wikipedia.com

Illustration 11..Page 101

　Creative Commons Share-Alike 3.0 License courtesy Wikipedia.com

Illustration 12..Page 101

　wikipedia.com "136-174MHz base station antenna" courtesy David Jordan

Illustration No. 13..Page 101

　Creative Commons Share-Alike 3.0 License

Illustration 14... Page 102

　Courtesy Staff Sgt. Matthew Smith; 4th Combat Camera Squadron; MacDill AFB, Tampa

　"Satellite Voice Antenna Installation—Operation Patriot Sands," 2013

THE AUTHOR

Michael A. Tewell worked briefly as a professional actor; a radio disk jockey and news director; and as a newspaper reporter—all before attending Kent State University at the age of twenty-five. At Kent State University, he received his Master of Arts in International Relations in 1979 and wrote his thesis on international terrorism (*Ideological Terrorism: An Analysis of the Capability and Will of Transnational Terrorist Groups to Attack the United States, 1979*).

He was admitted into Kent State Graduate College after his sophomore year and is one of the few college students to ever receive a master's degree without receiving a bachelor's degree. Tewell earned his Juris Doctorate degree from The National Law Center at The George Washington University in Washington, D.C. in 1982.

Rather that stay in Washington, D.C to work in national politics or the federal government, Tewell chose a legal career ensuring more direct contact with people's lives. He served twenty-five years as a trial lawyer defending the poor, the mentally ill, juveniles and veterans at the Office of Public Defender in Florida (Sixth Judicial Circuit).

In more than 30 years of practicing law, Tewell tried over 140 jury trials including homicides, manslaughter and other serious felonies and misdemeanors. He assisted in many other jury trials in a mentoring role and organized a trial practice clinic to teach new defense lawyers trial preparation techniques, pretrial motion practice, trial tactics and trial legal procedure needed to effectively represent clients at trial.

Since his retirement, Tewell enjoys life with his wife (also a criminal defense lawyer), their teenaged son, and a precocious cockapoo. His spare time is dedicated to writing on varied topics of interest, including spirituality, law, science and politics.

CPSIA information can be obtained
at www.ICGtesting.com
Printed in the USA
FFHW011523170219
50551130-55886FF